Basic
Construction Management
The Superintendent's Job

FOURTH EDITION

Leon Rogers

Home Builder Press®
National Association of Home Builders
1201 15th Street, NW
Washington, DC 20005-2800
(800) 223-2665
www.builderbooks.com

This publication is designed to provide accurate and authoritative information in regard to the subject matter covered. It is sold with the understanding that the publisher is not engaged in rendering legal, accounting, or other professional service. If legal advice or other expert assistance is required, the services of a competent professional person should be sought.

—From a declaration of Principles jointly adopted by a
Committee of the American Bar Association
and a Committee of Publishers and Associations

Basic Construction Management: The Superintendent's Job
ISBN 0-86718-452-3
© 1983, 1990, 1995, 1999 by Home Builder Press®
of the National Association of Home Builders
of the United States of America.
First edition 1983. Fourth edition 1999.

Cover by David Rhodes, Art Director, Home Builder Press.

Cover photos copyright © 1998 PhotoDisc, Inc.

Printed in the United States of America.

Library of Congress Cataloging-in-Publication Data
Rogers, Leon.
    Basic construction management, the superintendent's job /
Leon Rogers. —4th ed.
        p.   cm.
    Includes bibliographical references (p.).
    ISBN 0-86718-452-3
    1. Building—Superintendence. I. Title.
TH438.R612   1998   98-45915
690' .068—dc21   CIP

For further information, please contact—
    Home Builder Press®
    National Association of Home Builders
    1201 15th Street, NW
    Washington, DC 20005-2800
    (800) 223-2665
    www.builderbooks.com

12/98 Harlowe/Hutchison 3000
8/01 PA Hutchison/reprint 1500

Contents

Preface

This is the fourth edition of *Basic Construction Management: The Superintendent's Job*. The book has enjoyed great success and has been one of NAHB's best-selling titles. When the first edition was written in 1981 the world of construction was a very different place. The manuscript was typed five different times on a state-of-the-art IBM Selectric typewriter which cost more than a modern computer. Personal computers were unheard of, cellular phones had not been invented, construction scheduling was just in its infancy, Total Quality Management was a Japanese phenomenon, and construction management was largely "by the seat of your pants." We were just beginning to apply business management principles to residential construction projects. Sounds like the dark ages, doesn't it? But that was just a few years ago.

Much has happened in the past eighteen years. Most residential builders have computerized their operations to a great extent. Home buyers have become a lot more sophisticated and demand a lot more than they ever have in the past. Management of construction projects has become a lot more complex. Cost control and analysis of cost overruns and variances are now standard practices. Formal safety programs have become more common. Computerized scheduling is becoming widely accepted. Total Quality Management has been successfully implemented in many companies. Thanks to the educational efforts of the NAHB's Home Builders Institute, industry consultants, and builders who have continued to share ideas with other builders and educators in college construction management programs, today's builders are more highly educated and better prepared. But the competition is a lot tougher. Those who have survived in this industry have done so through better management and the application of some very basic principles.

Residential building today is an exciting and challenging business, one in which the construction superintendent plays a large and important role. Many demands are placed on today's superintendents. Construction typically includes greater customization of standard designs. Designs are more innovative and complex, resulting in more complex construction. Materials and methods are continually changing.

Superintendents now use computers on a daily basis for cost control, scheduling, and overall project management. Sometimes it seems as if the industry expects superintendents to be supermen and superwomen.

This edition of *Basic Construction Management: The Superintendent's Job* addresses these issues and discusses the latest developments in the management of home building operations. Special attention has been given to computerized scheduling and reporting systems, and to the importance of well-developed systems and systems management.

The feeling that comes from successfully organizing people, materials, and equipment to create a beautiful and functional home can be quite special. When you pass a home that you helped to build years earlier, you probably find it difficult not to look at it with at least a small sense of pride and think, "I built that!" This pride is at the heart of true success in the construction business. The truly successful are often not those who are wealthy or brilliant, but those who are genuinely good at what they do and who take pleasure in it.

Most people who are good at what they do apply simple rules and goals to their tasks. This book attempts to present such simple rules, targeted at maintaining your budget, complying with your schedule, and establishing quality control, leading to maximum profits—and an irreplaceable feeling of total success—in the long term.

About the Author

Leon Rogers is a professor of Construction Management at Brigham Young University (BYU). He also is the President of Construction Management Associates, a construction and consulting firm specializing in management systems for residential contractors. He has worked with a number of companies to help improve their management techniques and establish systems to monitor and improve construction operations. He has also developed a number of highly acclaimed employee training programs for construction companies.

One of the National Association of Home Builders' most popular national speakers, Leon Rogers has participated in more than 300 national seminars and has been an active member of NAHB for over 20 years.

In addition to *Basic Construction Management: The Superintendent's Job*, Leon Rogers is author of *Your Business Plan: How to Create It and How to Use It*, and *Production Management Template*, published by NAHB, and numerous articles on construction management.

Acknowledgments

The author extends special thanks to Lee Evans, The Lee Evans Group, to Wayne Homes, to Bob Whitten, to Prestige Home Builders, and to Stephen Covey for allowing excerpts and adaptations of their materials to appear in this book. Also thanks to Jerry Householder, chair of the Construction Department of Louisiana State University in Baton Rouge, whose contributions as co-author of the second edition of *Basic Construction Management* live on in the current book. And thanks to the many builders and industry experts who read and provided comments on the various outlines and manuscript drafts for each succeeding edition of this book. Reviewers for this fourth edition included: Jim Carr, Ohio State University, Wooster, Ohio; David DeLorenzo, NAHB Labor, Safety, and Health, Washington, D.C.; Allan Freedman, NAHB Builder Services, Washington, D.C.; Douglas M. Higgins, The Home Place, Inc., Jonesboro, Georgia; Mark O. Lords, The Snyder Group, Essex Junction, Vermont; Harold Swanson, Swanson Homes, Inc., Brookland Park, Minnesota; Michael Weiss, Weiss and Company, Inc., Carmel, Indiana; Allen Wells, American Home Place, Macon, Georgia; Richard Westlake, Hansen and Horn Group, Inc., Indianapolis, Indiana; Bob Whitten, Home Builders Advisory Service, Madison, Alabama; and William Young, NAHB Public Affairs, Washington, D.C.

Basic Construction Management: The Superintendent's Job, Fourth Edition, was produced under the general direction of Kent Colton, NAHB Executive Vice President and CEO, in association with NAHB staff members James E. Johnson, Jr., Staff Vice President, Information Services Division; Adrienne Ash, Assistant Staff Vice President, Publishing Services; Charlotte McKamy, Publisher; Kurt Lindblom, Project Manager; David Rhodes, Art Director; Sharon Lamberton, Copyeditor and Proofreader; and Thayer Long, Assistant Editor.

Hiring and Training Superintendents

Hiring the right people may be the single most important thing an employer does to promote business success. With remote production sites and a tremendous number of trade contractors, suppliers, inspectors, owners, and other people coming and going from each site, having the right people in place as superintendents certainly is critical to the success of a home building company. Providing timely and practical training for the new hires (and ongoing training for company veterans) may well be the builder's second most important contribution. Accordingly, this introductory chapter has been written for the builder (or hiring managers in larger building companies). Superintendents will also find the material of interest, however, as a thumbnail guide to the hiring process from the other side of the table—and as another way of looking at the superintendent's job.

Hiring a New Superintendent

Builders generally hire a new superintendent for one of two reasons: 1) the current superintendent has either left or is leaving the company, or 2) the company is growing and a new superintendent is needed to manage the additional work. In either case, depending on circumstances, the builder likely feels considerable pressure to find and hire someone quickly. Rushing a hiring decision is almost always a mistake, however. To make an effective hire, the builder must be prepared when a vacancy—or growth opportunity—arises.

The Hiring Process

Bob Whitten's excellent publication, *How to Hire and Supervise Subcontractors,* published by NAHB's Home Builder Press, provides detailed advice to guide you in making hiring decisions and following up with jobsite supervision of trade contractors. (For more information on this and other helpful publications, see Additional Resources at the back of this book.) The sections that follow summarize principles that can make the hiring process go more smoothly.

Growth Highlights Hiring and Training Needs

One builder found that hiring and training needs prevented his company from taking advantage of growth opportunities. As he put it, "I don't have the right people in place to allow me to grow. If I had the right people there is no doubt I could double my business. As it is, I will be lucky to grow 10 to 20 percent this year, and even that will probably bring us to our knees."

Another builder was able to take advantage of growth opportunities and opened several new offices. However, this company didn't have a training program in place so the next year became a real trial. Many homes took more than 200 days to build, the builder had some dissatisfied customers, and most projects were over budget. The company's overall profit margins were drastically affected. Fortunately, this builder was able to turn the situation around—but only after some intensive training and substantial turnover of production personnel.

Don't Wait Too Long to Start. Amazingly, many builders wait until the very last minute before deciding to hire somebody. As a result, that is exactly what they get—some *body*. The decision to hire should normally be made a minimum of three months before the new superintendent is really needed. Under the best of circumstances it takes at least three months to locate, interview, check references, re-interview, hire, and initially train a new superintendent.

A company may be faced with replacing a superintendent who quits with little or no notice. Even though you are in a tight situation, resist the urge to panic and hire the first vaguely qualified person you run across. The future of your company depends on the quality of your hiring decisions. Get the work done however you can while you are interviewing candidates for the position.

Network. In the hiring process "who you know, not what you know" definitely makes the difference. Whether as a job-seeker, a hiring manager, or an employer, you will be about twice as successful searching within a network of people you know than by advertising or contacting people off the street. Talk to your employees, trade contractors, suppliers, other builders or superintendents, remodelers, bookkeepers, people at church, friends, relatives—anyone you trust who may know of qualified people looking for work. Spread the word. Ask your current employees to keep an eye open. Develop a list of potential superintendents and update the list periodically to keep it current.

Look for College Graduates. Consider recent graduates from your local college or university. A number of very good construction management programs around the country annually graduate a ready pool of job candidates. Construction management graduates often are a little older and more mature than the typical college graduate, and they may already have several years of experience in construction. In addition, most construction management programs require an internship experience before graduation. Hiring interns offers your company a great opportunity to evaluate potential candidates for future positions at a relatively low cost.

The best time to recruit college graduates is in October for December graduates and in February for May or June graduates. The manager of student chapters for the

NAHB can provide you a list of colleges and universities that have construction management programs. The NAHB also holds a job fair in conjunction with its annual International Builders' Show and Convention. Hundreds of top students from throughout the nation attend the job fair and are available for interviews.

Some interns and college graduates have minimal construction experience, but they are computer literate and well educated. If your company has the resources and the time to train relatively inexperienced graduates, they can become valuable additions to your company. If your new hire must take on immediate, full construction management responsibility, you must carefully interview the candidates to determine their expertise.

Analyze Your Needs Before Interviewing. Develop or review your written superintendent's job description. Most builders find creating written position descriptions to be a very insightful experience. The exercise helps them solidify what they are looking for in their employees. Because written job descriptions can be such an effective tool, the Business Management Committee of NAHB has published a compilation of position descriptions for virtually every position in a construction company (see Additional Resources).

Consider the current and future needs of the company. Look at the career path for the new hire. How would each candidate likely fit in the organization over time? Look for candidates whose skills and personalities will complement those of your current employees. For example, if you already have people who are very good at building homes but are not good at customer relations, look for candidates with very strong interpersonal skills.

Use Applications to Develop a Candidate List. If you do not have an employee application form, develop one. You can obtain generic employee application forms at any good office supply store. Adapt the generic forms to your specific needs. Make sure that the form asks candidates to supply pertinent information, including a list of references. If possible, obtain a résumé from each applicant.

Rank the Applications. Once you have the applications and résumés in hand, read them carefully. Highlight the items that are most outstanding about each candidate and also any items about which you would like to know more. Then rank the applications from the most likely prospect to the least desirable. Discard all applications from candidates who are definitely unqualified.

Prepare a List of Questions. Before conducting any initial interviews, prepare a written list of questions you would like answered by each candidate. Use the same questions with all candidates; consistency is important.

When preparing the questions, you must decide what is important for you to know about each candidate. Consider asking open-ended questions, such as, "Can you give me an example in your previous employment when you were required to (name a specific task related to the position)?" Followup questions also are helpful, such as, "How did you react or handle the situation?" Pursue information that helps you discern how the candidate would perform in your work environment.

Don't rely only on first impressions, or how you personally feel about the candidate. An individual may be a great conversationalist but a poor organizer or a pushover as a superintendent. Ask questions that require the candidate to relate real-life experiences. For example, you could ask, "Can you give me an example of a time when you were required to hold the line with a trade contractor, and describe how you handled the situation?"

Asking job candidates to relate actual experiences from previous employment situations generally yields better results than does asking how they would respond to hypothetical situations. When relating actual experiences, candidates find it harder to guess what the interviewer is seeking. They must instead think of situations and relate what actually happened.

Other useful questions to ask include, "What is your greatest weakness?" "If you were hiring you for this position, what would be your greatest concern?" "What is your greatest strength?" or, "What do you have to offer in this position that no one else does?" You might also ask why the candidate is considering a change in employment or why the person is interested in the position at your company.

Don't Do All of the Talking. Inexperienced interviewers typically ask whatever question comes to mind and often spend the majority of the time talking instead of listening. After the interview they wish they had asked different or additional important questions. Research on interviewing techniques indicates that the most successful interviewers talk relatively little. An effective interviewer listens at least two-thirds of the time. Focus on asking a lot of leading or open-ended questions and then let the candidate express himself or herself.

Take Notes. Immediately after the interview take some time to jot down your reactions. Even if you have another interview scheduled immediately afterward, take the time to record your impressions. The few notes you jot down may be the most accurate information you will have. Do not take more notes during the interview than is absolutely necessary. Excessive note-taking may put the candidate on the defensive. Even a strong candidate may then become preoccupied with worrying about what they said or did wrong and wonder how they are doing.

Consider Using a Personality or Job Compatibility Profile. Many companies have found personality profiles and job compatibility profiles to be valuable screening tools. Such profiles can pinpoint with great accuracy the personality characteristics of a candidate. For example, if you need a superintendent who is organized and can work well under pressure without becoming frazzled, a personality profile can help you identify people who have strong skills handling stressful situations.

Personality and job compatibility profiles also provide a great deal of information that can be discussed in a second interview. You can simply ask the candidate to confirm whether and how the outstanding positive or negative characteristics indicated by the profile match the candidate's self-perceptions.

Check References. You are likely to obtain the best information about a candidate from people who know the candidate well. It is amazing how many people skip the important step of checking references, even though former employers can be excellent sources of information. The candidate will often supply a list of references. Expect these references to provide glowing remarks about the candidate. Ask such references hard and direct questions, and ask them for the names of other people who know the candidate well. Alternatively, ask for the names of two people who worked with the candidate at his or her last indicated place of employment.

In today's legal environment many people are reluctant to answer questions about job candidates. If you wish to obtain information beyond verification of salary or dates of employment, you must use patience and tact and develop trust with the reference before asking any tough questions.

Don't Settle for Just One. Don't settle for just one candidate or one interview. Above all, don't make an offer at the first interview. To ensure a complete assessment it is a good practice to interview each candidate several times in different settings. If your first interviews were held on a college campus (for example, at a job fair or as part of a recruiting day), bring the promising candidates to your operation for second interviews. Let the candidates see the environment in which they may be working. Give each candidate adequate time to ask you questions in the follow-up interviews. It is just as important for the candidate to feel good about you and your company as it is for you to find the right person for the job, and the candidates' questions of you may be as revealing as their answers to your questions.

Spend some time showing the candidate your operation. Explain how the position fits into your company. Introduce the candidate to the key players, especially those he or she would likely interact with most frequently. If possible, give the candidate time alone with those key players. Get the reaction of the key players concerning their interview(s) with the candidate. If the immediate supervisor is not doing the hiring directly, make sure the supervisor has the opportunity for an in-depth interview with the candidate. Spend as much time with the candidate in as many different ways as possible before making an offer. Remember, this person will make a long-lasting impact on your company's bottom line. You may want to involve two or more staff people in follow-up interviews. It is often easier to concentrate on the candidate when more than one person is doing the talking.

Re-rank the Candidates. Review your most important needs in hiring a new superintendent and then compare each candidate to the requirements. After the follow-up interviews rank the candidates. Note the strengths and weaknesses of each. Discuss the candidates with other managers and employees who were involved with interviewing.

Make a Final Decision. When you have weighed all of the data and searched your own feelings as well as the feelings of those directly involved, make a final decision. Consensus decision making is an important part of the process. Discuss openly the feelings of all concerned and then see if you can come to a consensus as to who should be hired.

Make the Offer. When you have made your final decision, prepare an offer for the best candidate. If you have done your homework you will have researched the market in your area and determined the competitive salaries or wages for superintendents. Determine the current salary and benefits requirements of the candidate as much as possible. Compare the candidate's requirements with what you are willing to offer. Prepare the offer including base salary (wage) and fringe benefits such as vehicle allowances, vacation, insurance coverage, sick leave, and so forth. Be prepared to answer any questions regarding the offer or benefits. When you make the offer, discuss the future career path and possibilities with the candidate; but in doing so, make it clear that their future career path is dependent on performance, market conditions, and the company's needs. Have the essential elements of the offer prepared in writing so the candidate can see them. Give the candidate a reasonable but limited time to think it over.

Train, Train, Train. Every employee deserves the opportunity to be properly trained. After making a hire, many employers forget the next—and perhaps most important—step: training. Often superintendents are hired "in the heat of the battle." You need them to hit the ground running immediately. But this is not likely to happen, so do not fool yourself or destroy an otherwise good employee. It does no good to spend a great deal of time and effort finding and hiring the right people only to burn them out in the first few weeks. Too-high expectations can cause new employees to become so frustrated that they quit.

Develop a good training program that presents the requirements of the job in a logical and controlled manner, without the pressure of ongoing problems and difficult situations. Teach new hires the way things should be done at your company. Teach them effective habits and practices. Often an employee will come to the job with some bad habits from a previous employer. Show him or her the right way to do things. Provide a good mentor for them. Supervisors or a well-qualified fellow superintendent will be invaluable in training the new superintendent. Develop a checklist of things the new employee needs to learn. Then develop exercises or opportunities the new employee can use in order to master them. As the new employee masters each item, check it off the list.

Have Fun. This may sound trivial, but it is very important. Have fun with the process. Finding and hiring new employees can be both very challenging and a rewarding experience. A lot depends on how you approach the task. The future of your company depends on the way you hire new superintendents. If you do it well, it can be one of the most exciting things you do.

Training a New Superintendent

The superintendent is one of the most important people in the entire organization. As new superintendents are hired, it is often critical to get them up and running quickly. Most builders simply have the new superintendents ride around with a veteran superintendent for a few days, then assign them a few homes to build. The number of homes steadily increases until the superintendent has a full load. Many new superintendents soon find themselves overwhelmed and burning out. A well-developed training program can give the superintendent a head start and allow him or her to come up to speed before being exposed to all of the pressures and problems associated with a full load of houses.

Many superintendents are hired with considerable experience working for other builders. The last thing you want is for them to bring bad habits learned in their previous employment into your organization. Every company has a "corporate culture" that needs to be learned; it is necessary to teach even experienced superintendents your company's policies and procedures. Training gives the new superintendents an opportunity to become familiar with your operating procedures and can help new hires fit in better and avoid mistakes.

One of the most important things that needs to take place in the training process is to introduce your new superintendent to the trade contractors and suppliers with whom he or she will be working. The most important responsibility of a superintendent may be to train trade contractors. To do this properly, superintendents must be properly trained themselves.

Training Process

The first step in developing a training program is to identify the various tasks for which a superintendent is responsible. In most construction companies the best way to develop a superintendent training program is to brainstorm together and come up with a tentative list of duties, or tasks, then refine the list until you are satisfied that it adequately describes the job of a superintendent. The tasks may already be gathered in a written job description (see Figure 1.1). Once the superintendent's job description has been defined, you are ready to outline the training required to bring specific individuals up to speed.

A four-step training process is ideal for new superintendents:

1. Observation: The new hire directly observes the trainer performing the operation or task (for example, a lot inspection or a homeowner walk-through).
2. Minor Participation: The new hire participates, with a minor role in performing the operation or task.
3. Major Participation: The new hire performs the operation or task, with the trainer present as a backup.
4. Evaluation: The new hire performs the operation or task, with the trainer present as an observer. The trainer evaluates the performance and completes a report to be placed in the training file.

Program Development and Implementation. Based on the job description, develop an outline listing the training requirements for a superintendent (Figure 1). Look at each task or group of tasks on the job description. Define the training required for each task. Assemble all of the documents, forms, and systems that a superintendent uses in the performance of his or her responsibilities. Describe the use of each document, form, or system and the role various parties play—especially that of the superintendent. Describe the interaction and communication needed for each task.

Develop activities and assignments to complement and evaluate the training. Quizzes can be used to evaluate comprehension of reading materials such as contracts or specifications. Exercises to strengthen superintendents' skills in given areas such as communication, negotiation, dealing with difficult customers, recruiting trade contractors, and so forth may involve role-playing or other practice. Assigned interviews with important people such as the company president, head of accounting, and customer service also facilitate learning about the corporate culture. Provide time to work in other areas of the company, for example, customer service, estimating, or bookkeeping to facilitate cross-training and better internal communication.

Assemble the rough draft of the training program into a prototype manual. Evaluate the materials and critique the results so far. Edit the manual and its contents to make sure it flows smoothly. Fine-tune the rough draft. Have a veteran superintendent review it and make comments.

Develop training aids that can be used in implementing the program. Overheads, exercises, narratives, and other training aids help maintain interest and add other dimensions to the training process.

Ongoing Training. Superintendents require continual training. Even veteran superintendents need to be reminded of their responsibilities. Many companies bring their superintendents together on a monthly basis for a day of training. Some of the topics suitable for an ongoing training program include the following:

- following company procedures
- completing paperwork and reports accurately and in a timely manner
- safety training
- training trade contractors
- other topics as needed

FIGURE 1   A Superintendent Training Program

A manual used in a construction superintendent training program might include the following sections and topics:

Introduction
- Program introduction
- Resource list
- Proficiency list (a checklist of all of the training requirements for superintendents)

Company Organization
- Company history
- Mission statement
- Organization chart
- Production organization chart
- Area map (a map of the area where the superintendent is expected to work; primarily a tool used by scattered-site builders)
- Organization assignment (an assignment, normally involving a series of interviews, developed to familiarize the new superintendent with the members of the organization)

Role of the Superintendent
- Section introduction
- Job descriptions for all production personnel
- Vehicle use and maintenance policy statement (outlining the use and care of company vehicles and reimbursement policies for use of personal vehicles)
- Tools and equipment list
- Dress code
- Cellular phone use policies (including suggestions for minimizing the high cost of cellular phones)

Sales
- Sales and marketing overview
- Contract(s)/purchase agreement(s) (an overview of the contracts or real estate purchase agreement(s) used by the company)
- "By owner" agreement (a review of company policy regarding work performed directly by the homeowner such as landscaping and paint)
- Project file (a review of all documentation that makes up the project file)
- Company plans and specifications (an exercise in reading and interpreting plans and specification)
- Homeowner package (a review of the various documents the homeowner signs and an exercise to help the superintendent understand the purposes and procedures relating to each document)
- Sales assignment (an exercise to help the superintendent understand what customers go through from the time they first make contact with the builder or visit the sales center until the home is actually started)

Site Meeting
- Site meeting definition (a meeting at which the superintendent lays out the house with the homeowner(s), establishes the grade, and locates the various utilities)
- Site meeting checklist (items to be covered at the site meeting)
- Site meeting assignment (initially, observation of a site meeting in which the new superintendent participates to a minor extent; later, the new superintendent takes full charge of a site meeting under the supervision of the trainer)

Preconstruction Meeting
- Preconstruction meeting (a conference at which the superintendent outlines the sequence of activities that will take place during the construction process and describes the responsibilities of the homeowner, builder and others during the process)
- Preconstruction meeting checklist (items to be covered in the preconstruction meeting)

Estimating and Purchasing
- Estimating and purchasing training (a review of the estimating and purchasing process, including all documentation)
- Estimate review checklist (items the superintendent reviews on the completed estimate to make sure nothing important was missed)
- Supplier and trade contractor list (an annotated list of trade contractors and suppliers currently working with the building company)
- Setup procedure for new suppliers or trade contractors (a review of the process for establishing a relationship with a new vendor or trade contractor)
- Quote sheets and price lists for suppliers and trade contractors (a review of the pricing structure used with the various trade contractors and suppliers)

Scheduling
- Section introduction (an introduction to scheduling systems and techniques used by the company)

(Continued)

FIGURE 1 *(Continued)*

- Updating schedules (a review of scheduling updating procedures used by the company)
- Scheduling training (exercises to teach scheduling to the new superintendent)

Construction

- Section introduction (a review of the construction processes and procedures used by the company)
- Communication (an overview of the need for effective, two-way communication between the superintendent and all parties involved in the construction process, including drafting personnel, estimating personnel, homeowners, and others)
- Communication assignments (an exercise or assignment to help the superintendent learn to apply better communication styles)
- Production manual (a manual containing the construction procedures and standards of the company; for a model, see NAHB's *Production Manual Template*)
- Quiz (a quiz to evaluate the superintendent's understanding of the items and information included in the production manual)

Quality Control Checklists

- Checklists (trade-specific lists developed as tools to assist superintendents in evaluating the quality of work performed by trade contractors)
- Quality control assignments (a series of inspections to be performed at the various levels by the new superintendent)

Managing Trade Contractors

- Trade contractor agreement (a review of the subcontract agreements used by the company)

- Workers' compensation (a review of requirements and certificates required by the company)
- Keys to working with trade contractors (an overview of management principles as applied to relationships with trade contractors)

Health and Safety

- Safety and health training (OSHA and safety and health training)
- Quiz (a quiz to evaluate the superintendent's understanding of OSHA and safety and health requirements)
- HazCom and MSDS training
- Safety inspection assignments (the new superintendent participates in, then performs jobsite safety inspections)

Corporate

- Additional training
- Drafting training
- Accounting and job cost training

Customer Service

- Limited warranty exercise (a review of the documentation with an exercise to help the superintendent understand the warranty used by the company)
- Customer service training
- Customer service exercises (the superintendent works with customer service personnel on warranty calls to reinforce the importance of quality control and scope of liability)

1

The Superintendent's Job

The superintendent's job is arguably the most critical position in any residential construction company. Whether your company builds two custom homes a year or a thousand production homes, your primary business is the construction of homes, and as the superintendent you have more control over the building operations than anyone else does. In the field, there is little room for error. As one CEO said, "It doesn't matter how many homes we sell; if we can't build them under control, the rest doesn't matter." The success of the organization depends largely on your ability to manage your projects. This book has been written specifically for you, the residential construction superintendent responsible for seeing that all fieldwork is performed properly according to plan.

Your specific responsibilities within the company may vary. In a large company, the superintendent usually reports to a project manager, who then reports on up the management line until the owner or chief executive is involved. In smaller construction companies, the superintendent, project manager, and owner are often the same person. Whether you are the company owner or several management layers away from the executive office, your number one goal as a superintendent remains the same:

Maximize profits in the long term while maintaining a standard of excellence within the homes you produce.

The profit of a construction company is the difference between the sales price and the cost of each house, including overhead. Superintendents ordinarily have only limited control over the ultimate sales price of the homes they build. Through quality workmanship and orderly jobsites, however, they exercise their greatest influence over company profits by controlling cost. Through attention to detail and good management of trade contractors, in-house labor, and materials, you can influence the final cost of any project significantly. While your influence on project cost is relatively easy to understand, your impact on the company's continuing profits in the long term may be less clear.

For example, the general philosophy of most businesses is to make the greatest possible profit for the period of time in which the company is in business. If a builder plans to build only one house and then immediately go out of business, then a short-sighted, "quick-buck" approach might yield the highest profit. Fortunately, most construction companies plan to stay in business for a long time. As a superintendent, your key to success is motivating your workers and trade contractors and suppliers to bring in every project:

- on time
- within budget
- according to established quality standards

Just about every decision you make as a superintendent should be made in keeping with these three basic responsibilities. Of course some aspects of your job may be aimed toward achieving goals that don't fall directly under one of these categories. For example you may need to oversee maintaining a safe work environment. Unsafe projects generally result in accidents that cause delays and cost a tremendous amount of money. So even safety affects the bottom line. By and large, however, your energies should always be geared toward bringing in a quality job, safely, on time, and within budget.

While this manual provides many techniques and examples with specific applications, it is not intended to teach a particular system or point of view. The purpose of this book is to teach you correct principles and concepts and provide specific examples to help you develop a system of construction supervision that will result in improved performance and maximum profits for your building company.

The Superintendent's Role

A superintendent's responsibilities vary greatly with the size of the company and the leadership strategy of its owners. In a large construction company, your responsibilities as a superintendent might be limited to a single jobsite, or even to a single segment of the job, such as framing or concrete work. In the majority of residential construction companies, however, your role will be much more complicated, involving the entire production phase of several homes at the same time. Therefore, you must have knowledge of every aspect of construction and be familiar with the work of every trade.

In construction management seminars across the country, builders often ask, "How many jobs should a superintendent be responsible for at one time?" The answer to this question depends on a number of factors. How many homes a year do you build? How similar are the homes? Superintendents who build from a defined set of standard plans and handle little, if any, customization should be able to manage many more homes at one time than those who build custom homes. How much owner involvement is allowed? Building on the owner's lot and according to his or her plans is entirely different from building speculative homes in a confined subdivision with no changes. Other factors determining a superintendent's role include the size and complexity of the homes, the level of trade contractor involvement versus in-house labor, the architect's involvement in construction, the proximity of jobsites in relation to each other, and a host of other variables.

Some generalizations may be helpful, but remember that each situation is unique. If you build small- to moderate-sized homes in a particular subdivision according to standard plans, you may be able to manage 20 to 30 homes at a time. If you build one home at a time on scattered sites, you may be able to handle 10 to 12 at a time. If you build large, totally custom homes, you may be able to manage only one to two homes at a time. Lee Evans, considered by many to be the most widely recognized consultant in the residential construction industry, suggests another way to look at the question. According to Evans, the cost of supervision should be about $1\frac{1}{2}$ percent of sales. If your sales-to-supervision ratio is less than 1 percent, the superintendent may be overworked. If the ratio is greater than 2 percent of sales, the superintendent may be underused.

The Superintendent's Authority

The amount of authority you have as a superintendent depends upon several factors, including the size and structure of your company, the scope of the job, and your level of experience. It is essential that you fully understand the extent of your role and authority within your particular organization.

Most building companies define a superintendent's duties and authority formally by means of a job description. Job descriptions specifically define the extent of the authority of each position within the company and the interaction between the various positions in the company.

Smaller or newly formed companies may be less rigid or well defined in their organization. Such companies generally offer a superintendent greater freedom. Most builders simply want the job done right and expect the superintendent to take over and supervise the entire construction process. Whether the company is small or large, most builders are looking for self-motivated, take-charge supervisors who project this image to others. One of the keys to success is defining the role of each of the participants in the construction process.

The Superintendent as the Company's Agent

A superintendent is a construction company's agent and field representative, and unless specifically limited otherwise all negotiations, agreements, and contracts the superintendent enters into become legal and binding. You therefore have a moral and legal obligation to represent a building company's position responsibly and properly. While you may not agree with all decisions made, it is your duty to put the company first and see that all established goals and objectives are achieved. Superintendents must not allow their personal interests to come between the interests of the company and other individuals such as trade contractors or clients.

As a superintendent, you direct—and often hire and fire—most construction employees and trade contractors and usually have more direct contact with people, both inside and outside the organization, than anyone else in a building company. In addition, you may often schedule trade contractors, order materials, inspect to ensure quality work, warrant the home, and even deal directly with customers. You can substantially improve a company's image and reputation by being an effective superintendent.

The Superintendent's Duties

As mentioned earlier, superintendents have responsibilities in nearly every facet of the construction business. They interact with just about everyone within the company. Every company should use job descriptions to establish the duties and responsibilities of the superintendent and the lines of authority. A job description should define and document your primary responsibilities and establish a basis for performance evaluation. Job descriptions should also be specific about the relative amount of time and attention that should be devoted to each duty. Being aware of how your time is spent is an important management practice. One of the easiest ways to do this is to monitor the amount of time you spend on each task over a period of time and then make adjustments as necessary in order to improve your effectiveness as a manager. For example, if you spend too much time stamping out fires, it is a good indication that you need to spend more time planning in order to avoid future crises. An example of a very detailed job description is shown in Figure 1.1. Your job description may not be as detailed as this example, which is given primarily to bring to your attention the extent of the responsibilities and duties of a typical superintendent.

The Superintendent as Leader

As a key member of a builder's management team, you must also add leadership to your list of superintendent responsibilities. Leadership is the ability to work with and get things done through others while winning their respect, confidence, loyalty, and cooperation. Many people still believe that leaders are born, not made. In reality, leadership is an attribute that can be acquired and developed by anyone with the necessary motivation.

Leadership Basics

One of the keys to being a successful superintendent is understanding your role as a manager and as a leader. In his best-selling book, *Principle-Centered Leadership*, Steven Covey stresses that the distinction between management and leadership is a crucial one. According to Covey, most people in business spend too much time managing and not nearly enough time leading. We are far less effective when we try to manage people. We should manage things and lead people. In other words, you manage all of the other resources at your disposal—budgets, equipment, materials, and other items. However, the human resource—your most important resource—cannot be managed; people must be led. This includes in-house construction workers, trade contractors, suppliers, and others. The old "bull-of-the-woods" superintendent is a thing of the past. You can't force people to perform. Intimidation gets you nowhere. When traditional superintendents try to force their will upon today's workers and independent trade contractors, many of them will rebel and refuse to work, or worse, they will subvert the efforts of the superintendent. Therefore, as an effective leader, you must set an example of cooperation and good will, coupled with discipline, to get the maximum cooperation from your subordinates. The adage "firm but fair" still works with most people. It is much easier to listen, learn, and cooperate than it is to shout, be obstinate, and point fingers.

In order to obtain the best performance from your people, including trade contractors, allow them a considerable amount of discussion and input in the decision-making process. Of course, responsibility for any final decision rests with you, the superintendent, the leader. However, enlisting the cooperation of those who will actually perform the work, recognizing their input, and understanding the points they make can lead to valuable support and insights.

With more and more building companies subcontracting almost all of their work, superintendents increasingly work directly with several different, completely independent businesspersons. The more independent these trade contractors are, the more cooperation and coordination is required. If you have a trade contractor who

FIGURE 1.1 The Superintendent's Duties

Reports to: Production Manager

Objectives: The superintendent oversees construction of individual homes to ensure high quality work that is produced under budget, on schedule, and to the homeowners' satisfaction. The superintendent takes full responsibility for producing homes in an efficient and safe manner through effective management of trade contractors and suppliers, and under the direction of the production manager directs the activities of trade contractors, material suppliers, inspectors, and others on all homes under his or her direction. The superintendent is the primary representative of the company on construction matters to homeowners. He or she orchestrates the work, coordinates the various complex aspects of the construction process, and trains trade contractors as needed in the performance of their work.

Responsibilities: The superintendent must demonstrate the following general abilities, qualities, and knowledge, and perform the following tasks:

- maintain a high degree if integrity and honesty in all business dealings
- be self-motivated and solve problems effectively
- communicate well; know how to listen. "Seek first to understand, then to be understood." (Stephen Covey)
- be well organized and thorough
- get along well with others; be empathetic and lead by example
- drive courteously, conscientiously, and safely, and maintain a positive driving record
- be emotionally stable and able to work through difficult situations in a calm and professional manner
- be professional in dress, manners, and conduct; represent company at all times and in all places, and maintain a high degree of integrity and company loyalty
- be able to read, understand, and resolve inconsistencies and problems in plans and specifications
- know applicable building codes, zoning ordinances, OSHA requirements and other legal restrictions
- maintain a current record of all code interpretations and local ordinances for each jurisdiction in which he or she works
- know and understand the performance criteria and construction standards of each of the trade contractor trades
- know safety practices and procedures and conduct an effective overall jobsite safety program
- understand basic surveying principles and practices and interpret site-specific topography

- understand various scheduling theories and methods
- maintain a current, accurate understanding of accounting and other business practices
- solve problems within his or her authority
- delegate responsibilities to subordinates while remaining responsible for their performance
- make quick, accurate decisions when necessary and take responsibility for decisions
- review business forms, checklists, and reports that aid in controlling aspects of the construction process under his or her responsibility
- be familiar with and use established supervisory and motivational techniques to elicit peak performance from all employees, suppliers, and trade contractors in relation to quantity as well as quality
- be computer literate and know how to operate word processing, spreadsheets, database, and applicable scheduling programs
- complete and submit all necessary production reports and information in a readable, accurate, and timely manner
- work with the appropriate manager to establish pricing for custom options
- maintain a list of available local trade contractors that strikes the company's desired balance of pricing and quality
- be directly involved and keep abreast of the latest construction industry trends through study of literature, talking with material suppliers, and so forth
- understand the need for uniformity in construction methods within the company, support company policies and construction standards, suggest improvements within the system, and execute policies to ensure compliance with company quality standards
- understand and approve the basic materials and methods used by trade contractors and assure compliance with company construction standards
- attend seminars and training sessions on production-related subjects
- attend superintendents' training meetings
- plan, organize, and conduct ongoing training sessions for construction personnel and with others related to the trades
- work closely with the salesperson to review initial job file
- make the transition from sales to production a pleasant experience for the homeowner(s)

(Continued)

FIGURE 1.1 *(Continued)*

- build high quality homes through the use of employees, suppliers, and trade contractors
- achieve high quality work and homeowner satisfaction through effective management of resources and the construction process
- consult with engineers, architects, and designers on design problems
- maintain effective, cooperative, working relationships with architects, engineers, trade contractors, employees, material suppliers, homeowners, public officials, and the general public
- attend scheduled production meetings, training seminars, staff meetings, and other meetings with production manager and other staff
- determine alternative work assignments during inclement weather or when schedules are changed
- purchase (with proper authority) and oversee maintenance and care of company-owned hand tools and equipment
- evaluate alternative methods of construction for cost efficiency and improved quality; propose necessary changes to production manager
- instruct, train, and work with trade contractors and others to assure compliance with proper, approved methods of construction and operating procedures
- work with new superintendents to properly train them; act as a mentor to new superintendents
- monitor quality control standards and supervise their implementation
- work directly with office personnel (estimating, drafting, accounting, and so forth) and with others as needed to develop or modify company systems to ensure a smooth production operation
- organize trade contractor work force and schedule their work in cooperation with the production assistant
- conduct site meetings and preconstruction meetings with homeowners and review plans
- verify accuracy of maps in relation to accuracy of direction, mileage, and so forth; correct maps as necessary
- lay out the home on the lot, establish elevations and depth of excavation, and adjust the plans to accommodate specific site and code requirements

- erect job signs, trash signs, and job boxes on all construction sites
- communicate site changes to drafting and estimating personnel accurately so that the plans and estimate can be corrected and reflect accurate quantities for homes under the superintendent's direct control
- review estimates of materials and labor needed for the construction process
- establish and enforce safety measures; make sure trade contractors perform their work safely and in accordance with OSHA guidelines and that trade contractors have an effective safety program
- be intolerant of unsafe work practices and take appropriate but firm action to promptly remedy safety violations
- submit trade contractor information (price list, trade contractor agreement, checklists , and so forth) to the accounting department so accounting records can be kept current and accurate
- be personally present on the job when excavation begins to ensure that the proper grades and depths of excavation are maintained
- prepare or review weekly production reports and other construction performance records
- submit accurate information and paperwork in a timely manner so that managerial reports can be generated
- make sure all construction schedules are accurate; update schedules daily
- use two-week interval schedules to monitor and manage construction of homes
- review project status reports weekly or as required by company policy; work with peers to coordinate scheduling of common trade contractors
- monitor construction times and focus attention of all construction personnel to ensure a smooth, efficient, and continuous flow of work on each individual home
- make sure homeowner change orders are completed, signed, and approved before the change is made
- walk homes on a daily basis
- review quality checklists with all trade contractors at appropriate times

(Continued)

FIGURE 1.1 *(Continued)*

- conduct detailed inspection of each phase of construction before any work is authorized for payment; use quality checklists for inspections, quality control, and standard operating procedures

- work to develop and improve a cooperative, working relationship between local building inspectors, city and county agencies, and other members of the building team

- conduct a detailed framing inspection with the framing trade contractor before completion of framing work

- conduct pre-drywall inspection with the homeowner(s) and ensure all structural, mechanical, and electrical components are in the exact locations

- review purchase orders and monitor authorization of payments to trade contractors and material suppliers by the accounting department; make sure trade contractors' work is 100 percent complete before authorizing payment

- write variance purchase orders (VPOs) daily for all items that were not included in the original purchase order and accurately justify the reason for each variance; fax VPOs to the corporate office daily

- give initial approval to trade contractors and suppliers of purchase orders, VPOs, and homeowner service work orders; monitor authorization of payments to trade contractors and material suppliers by the accounting department

- coordinate the return of all excess or inferior material; make sure credits are filled out and send necessary paperwork to estimating

- maintain a clean and safe jobsite during construction, including site conditions and streets; ensure that neighborhood streets and yards are clean and that trash is cleaned up; make sure silt control fences are in place and serving their purpose

- develop positive customer relations with each homeowner based on timely performance; ensure complete customer satisfaction by meeting regularly with homeowner(s) and maintaining effective two-way communication with them

- communicate with each homeowner at least once a week, more often as necessary

- keep a log of all communications with homeowner(s); answer questions and address homeowner concerns during construction and whenever necessary

- coordinate installation of utilities and final connections with homeowner(s), local and code officials, and utility companies

- handle negative situations and homeowner concerns in a diplomatic and professional manner while trying to maintain excellent relations with homeowner(s); know when to say "no" and how to do so a way that does not offend the homeowner

- keep homeowner informed of the status of all allowances

- coordinate the work of all trade contractors on the jobsite to avoid conflicts and dry runs

- walk each completed home before the homeowner walk-through to make sure that the home is complete and clean and meets standards of quality

- make sure all punch-list items are completed quickly and professionally (within one week maximum); make necessary repairs properly, avoiding quick fixes or coverups

- make sure homeowners sign off before settlement (no punch-list items are pending)

- make sure home is 100 percent complete and meets standards of excellence

- participate in homeowner walk-through as necessary; ensure a smooth "hand off" of the home to the quality personnel and that the homeowner is happy with the transition

- review and verify accuracy of allowance records with accounting before homeowner settlement

- make sure all final bills have been paid

- work with accounting department to perform an analysis of each house after all bills have been paid

- assist the customer service or warranty personnel and trade contractors on warranty service and homeowner complaints

- help maintain the construction office and keep it clean and organized

- integrate all of the above tasks into an organized, controlled, smooth-flowing system

- accomplish any other tasks as required by the position or the production manager

works regularly or exclusively for you, you still have a tremendous amount of direct control. On the other hand, if you are a small-volume builder who uses trade contractors only on occasion or if you award each subcontract independently based on the lowest bid, you will have a continual training and coordination problem. Developing relations with key trade contractors in each trade and using them exclusively is a wise practice. It is critical that both the superintendent and trade contractor be clear about what each expects of the other in the working relationship. Dialogue and documentation are important in matching expectations.

Leadership Styles

Different leadership styles are characterized by the particular emphasis each places on the decision-making process. At one extreme is company-centered leadership, in which the desires and needs of the company come first. At the other extreme is the subordinate-centered leadership style, which puts the needs of employees first. The most effective leaders maintain a balance between the needs of the building company and those of the individual worker. Maintaining this balance often requires a dynamic, forceful, and flexible individual. You may find that it helps to vary your leadership style to best handle each particular situation. This process often results in the establishment of leadership precedents—choosing an approach to accomplish the objectives involved in one situation and using that style for similar situations. Leadership styles often used in the construction industry are discussed in the following sections.

Autocratic. In this traditional leadership style, also referred to as dictatorial leadership, superintendents keep all authority themselves and delegate little, if any, to subordinates. The key phrase is "total control." The superintendent gives complete direction and makes all decisions; discussion and suggestions are at best tolerated but not encouraged. To employ this leadership style successfully, you must become a forceful leader of people and have a high degree of proficiency in the each of the crafts and skills you supervise. This leadership style is most appropriate when:

- an emergency arises and there is no time for discussion
- an extremely tight schedule must be met
- an employee directly challenges your authority as a superintendent
- an employee behaves stubbornly or proves difficult to work with
- a new employee is being trained

Bureaucratic. This leadership style relies heavily on established rules, regulations, policies, and procedures to govern the organization. The leader simply enforces the various regulations. The bureaucratic approach works best when:

- activities must be performed according to a strict plan
- highly technical equipment must be installed
- specific procedures (internal or external) must be followed
- critical safety issues are involved
- new employees are being trained
- new policies must be implemented

Democratic. The democratic or team approach to leadership allows subordinates to participate in company goal setting. Many employees today are motivated and competent enough to work well under democratic leadership. This leadership style works best when:

- workers are well educated, experienced, and motivated
- schedules include time for subordinates' participation in decision making
- problems require long-term solutions
- grievances need to be aired and tensions relieved

A superintendent may use a democratic leadership approach when, for example, a company is seeking to improve the compensation or bonus system for employees. In this situation, when new policies will directly affect employees, the effective superintendent does a lot of listening and little talking.

Orchestrative. An orchestrative leader relates to employees in much the same way a conductor relates to the members of an orchestra. For this leadership style to be effective, all workers must be highly skilled at their jobs and motivated to work and succeed independently. They must also possess pride in their work. Highly competent and independent workers tend to prefer this leadership style. Orchestration is particularly suited to situations in which:

- skilled, experienced personnel can meet their responsibilities with complete confidence
- established policies and procedures allow you to comfortably delegate tasks and responsibilities
- problems can be solved through team efforts
- a new supervisor who lacks experience in the company's day-to-day operations can gain from the experience and expertise of workers

A good time to call upon the orchestrative leadership approach is during the peak of home production each year. During this hectic time when resources are spread thin, a superintendent may delegate some tasks, relying on the good judgement of employees to get through the busiest weeks of the year.

What's in a Name

The traditional title for managers of the construction process has been *superintendent*. In the past few years, as the responsibilities of the superintendent have increased, many builders have considered a change in this title. Some builders use the title *construction manager* while others use the terms *project manager, construction supervisor,* or even *builder.* Many believe that the term *superintendent* simply does not convey to higher level managers or to clients the full responsibilities of a complete manager of the construction process. An appropriate title must reflect the complexity and requirements of the position. In addition, the title projects an image to customers, trade contractors, suppliers, inspectors, and others.

Builders should consider what image they really want to project to those who interact with their superintendents. A few companies use the title *builder* for their superintendents because they want homeowners to feel that the person in charge of their home is a builder who has full responsibility and authority. They want to facilitate development of a bond or a personal relationship between the builder (superintendent) and the client; a relationship where the customer is proud to say, "This is my builder." This choice seems to work very well for some building companies.

Another useful title is *construction manager.* This title conveys the fullness of the responsibility carried by the superintendent. It also conveys to the customer, trade contractors, and others outside the organization that this is the person who is in charge. The construction manager, in turn, takes additional pride in his or her position and level of responsibility.

The Superintendent as Manager

As a superintendent, you are part of a management team. Therefore, you should be trained as a manager and supervisor and act primarily in that capacity. In too many companies, superintendents brought up through the trades are placed in supervisory positions without even the most basic management training. This lack of training can prove detrimental when novice superintendents suddenly find themselves overseeing production and quality control rather than producing the work themselves. Your performance as a manager is measured by what others do. How well you motivate workers and trade contractors to produce is a key factor in your success.

People who study management often divide the topic into four basic elements: planning, organizing, directing, and controlling. The effective superintendent successfully applies all four of these elements to reach the three goals of bringing the job in on time, within the budget, and according to established quality standards. By understanding these basic principles and applying them on your jobs, you can become a better superintendent.

Planning

Someone once summarized management planning in the following way:

Plan your work, and work your plan.

Planning is the most basic and most important function of management, requiring that you simply figure out what must be done. When you plan, you develop a program of action to achieve stated goals through the use of people, materials, and financial resources. While, as a superintendent, you are responsible for planning all the activities that come under your direction, you also must implement the plans of the builder to whom you report. The saying, "Proper planning prevents poor performance" holds especially true in the building industry. Problems arise when no overall objectives are established, when policies and programs conflict, or when procedures are wasteful or poorly thought out.

Many superintendents understand the need for planning but have never been taught what or how to plan. Bob Whitten, a builder, industry consultant, and former director of business management at the National Association of Home Builders (NAHB), has developed an outline to guide superintendents in planning. With his permission, I have adapted the outline for use in this manual.

Planning is a continuous process, but certain planning activities should be scheduled or reviewed regularly on a yearly, quarterly, monthly, weekly, or daily basis.

Yearly Planning. On a yearly basis, the superintendent should review, evaluate, and revise, as needed, the following:

- subcontract agreements
- performance of in-house construction personnel
- projections of staffing and training needs
- scheduling system
- quality checklists
- communication systems (internal and external)
- annual forecast of sales, starts, and closings
- production policies and procedures
- managerial policies and procedures

Quarterly Planning. On a quarterly basis, the superintendent should monitor the status of and adjust, as needed, the following:

- all yearly items
- quarterly forecast of sales, starts, and closings
- performance and availability of trade contractors (to establish a stable trade contractor base)
- potential material shortages and price increases
- accuracy of construction projections and schedules
- changes to production manuals

Monthly Planning. Monthly, according to projected workload, the superintendent should analyze and take steps to adjust, as needed, the following:

- trade contractor availability and scheduling
- recurring material and production problems
- project status reports
- sales reports
- housing starts for next two months
- closings for the month

Weekly Planning. Weekly, unit-by-unit, the superintendent should evaluate and adjust, as needed, the following:

- long-range weather forecast
- jobsite visits and overall status of all active jobs
- starts and closings
- communications with homeowners
- meetings
- trade contractor schedules for the next three weeks
- delivery schedules
- inspection schedules
- critical places to be at specific times of the week

The superintendent must also communicate revised schedules to all concerned parties and analyze the long-range weather forecast to anticipate factors that may interfere with the schedules.

Daily Planning. Daily, unit-by-unit, the superintendent should review, verify, and adjust, as needed, the following:

- short-range weather forecast
- work to be accomplished today, schedule updates
- worksite readiness for next scheduled trade contractor
- material drop zones and sites (to be sure they are properly prepared and marked)
- daily communications with homeowners
- daily communications with trade contractors and inspectors (including coordination and scheduling for near-future work)

The superintendent also must discuss relevant problems and corrective measures with the builder or other supervisors, communicate any last-minute schedule

The "To Do" List

Trying to carry around a list of half a dozen tasks in your head takes up valuable storage space in your memory. One of the easiest yet most effective methods of planning involves maintaining a written list of all things to be done each day. In the past ten years there has been a proliferation of time management systems, some manual and others computerized. These systems help you keep track of appointments as well as the things you need to accomplish each day. Many of the new computerized scheduling programs include a built-in, updatable list of tasks that need to be completed. Each day, you can generate and print a new list of critical tasks. The list is integrated with your activities on the schedule. For example, a day or so before you are scheduled to place the footings, the program reminds you to call in the inspection and place a "will call" for the concrete. Any items you don't complete are automatically carried over to the next day's list. These programs are marvelous management tools for the superintendent.

Your time-management system may be as simple as a $15 day planner purchased at an office supply store or a $35 electronic pocket organizer; or it may be as complex as a $4,000 fully integrated scheduling software package for your personal computer.

changes, and communicate with the homeowners if necessary. Because so many quick decisions occur each day on the jobsite, many superintendents use a daily "to do" list to help ensure that no critical decision or action is forgotten. Learning to prioritize tasks and handle unforeseen problems according to their impact is an important aspect of daily supervision.

Organizing

The tasks of planning and organizing work often overlap. Once you have determined what needs to be done during the planning stage, you must organize the job, determining who is going to perform the work and how completion of the task fits into the overall scheme.

The organization of work boils down to one question: *Who* does *what* by *when?* This simple formula is the key to getting things done. The "who" part of the question requires you to allocate personnel and other resources to accomplish the task. It also involves the establishment or use of lines of authority and responsibility to guide the assigned workers. "What" requires you to determine exactly which tasks will be performed. Finally, "when"—the phase of organization most often overlooked—must be carefully planned, including a factor to accommodate what can go wrong. Chapter 5 provides detailed information about how to schedule activities on the jobsite for maximum efficiency.

You must consider two additional organizational factors. The first is your own authority as a superintendent to organize workers to perform a certain task. The second is a clearly understood policy controlling hiring and firing, which will allow you to maximize your organizational efforts. For example, do you as a superintendent have the authority to fire a trade contractor for failure to comply with company policies or poor quality of work, or is this decision reserved for the company owner? In most cases specific criteria will govern firing of trade contractors.

Directing and Coordinating

The majority of a superintendent's time is spent directing and coordinating tasks. Therefore, these efforts must be targeted directly at the primary goal. However, through better planning and organizing, you can be much more effective and get the most out of your limited time. If planning is complete and your objectives are reasonable, all activities should contribute to achieving your goals, and any deviations can be corrected through effective control procedures. You should spend most of your time directing and coordinating those areas critical to company success. If a project progresses according to the plan, no additional directions are needed. As much as possible, handle routine matters through routine policies and procedures. Only extraordinary items falling beyond preset standards should require management attention. Of course, any deviations important to upper management should be reported to the builder or your superior in a daily or weekly report.

Controlling

Superintendents are directly involved in controlling activities. The scope of your activities may include control of materials, construction methods, labor, waste prevention, and costs. In addition, you will ordinarily have the authority to make decisions that affect the selection of trade contractors and the quantities and cost of labor used. In short, your job is to control events to conform to the plan you have already established, note any variances from the plan, and develop alternative plans to mitigate damages and get things back in line.

Superintendent Control System and Reports. Your entire control system should be economical—detailed enough to highlight deviations from your plans, yet not so detailed as to become burdensome or overly expensive. While complex controls can be extremely useful, they can become so complicated that they are rarely or never used. Therefore, your reports should be timely, but not time-consuming. Always keep in mind that management needs to keep up to date on job progress in order to make accurate, sound decisions. It will normally take about an hour each day to fill out and update the various reports required. Some common reports include the following:

- daily "to do" reports
- updated schedules

How Much Do You Need To Know?

How much privileged information should be given to superintendents? Builders have yet to agree on how much information to give superintendents. Some feel that much sensitive information should be held from superintendents. Pricing and cost information, for example, often are held in the strictest confidence. Builders who withhold this information avoid the risk that it will "leak out" and somehow be divulged to the competition.

Others feel that superintendents should be privy to almost all information, including financial information. They contend that the superintendent is a key person within the organization and, as such, needs the best and most accurate information available in order to manage effectively.

As a superintendent for two builders, I built well over 100 homes without being given any information concerning cost or even quantity of material with which to build them. Presumably, the builder prepared an accurate estimate for the home as part of the bidding process. However, that information was withheld from superintendents. To adequately perform the job and contain costs, a superintendent needs to be aware of the expenditure of funds on each project in relation to the original budget. Superintendents need to have all of the information necessary in order to build the home, including cost information. In return, the builder must be able to trust that the superintendent will keep sensitive information confidential.

- cost control variance reports
- customer service reports
- project management progress reports

Project management reports typically summarize the status of each job and any commitments made, usually in the form of a daily log.

Accounting and Cost Control. Superintendents are also directly involved in controlling the cost of each construction project. In order to accurately track and report the financial status of each job, accounting and bookkeeping personnel must be kept informed of cost-related activities. The superintendent's daily progress reports will provide much of this information.

Accounting departments in larger companies may have their own forms that require you to fill in specific information. Whether you are filling in forms for an accounting department or simply handing progress reports to your builder, be careful to safeguard all financial information, sharing it only on a "need-to-know" basis and using it wisely.

The superintendent must carefully review supplier and trade contractor invoices and payments, purchase orders (before and after deliveries and work completion), and work orders. Work must be completed satisfactorily before authorization of payment. You should have a basic understanding of lien laws as they pertain to a builder's financial liability in order to prevent possible legal problems with trade contractors.

Construction Activities

A large part of your responsibilities as a superintendent will involve the coordination of the basic elements of construction: materials, labor, equipment, trade contractors, customers, financial resources, and time. Each of these elements affects and is affected by the others. To coordinate these elements, you will need to develop a well-organized system that incorporates four essential activities: estimating, scheduling, staffing, and evaluation.

Estimating

Depending upon the company's size and organizational structure, estimating may be performed by staff estimators, the builder, the superintendent, or someone else in the company. Superintendents may not be asked to estimate jobs, but they are expected to build each job according to its estimate. Because you probably know more about the actual construction process than anyone else involved, you can and should be an information source for the estimator, providing accurate, reliable information that will decrease the chances of estimating errors. At bare minimum, the superintendent should review the estimate to make sure it reflects the materials and methods that will be used in the construction of the project. Review of a typical estimate normally requires at least two to three hours, and complex jobs may require much longer. Chapter 2 provides a more detailed list of key items the superintendent must review.

Scheduling

Scheduling is one of the primary duties of the superintendent. Together with the builder, the superintendent should select a scheduling system appropriate for the company. You may use the critical path method (CPM), a bar chart, a time line, or some other form of scheduling. As the superintendent, you will ultimately be responsible for employee and trade contractor deadlines; therefore you should participate actively in determining who does what, in which order, and when. It is your responsibility to evaluate alternatives, establish contingency plans, and make sure that employees and trade contractors are scheduled in an optimum manner. The overall schedule needs to be developed within the guidelines of your company policies.

Developing a schedule for an average house normally requires from one to four hours, depending on the size and complexity of the house and the level of detail in the schedule. The scheduling method you use also depends to a large degree on the number of concurrent projects you will supervise. For example, if you are responsible for multiple projects, you are more likely to make use of computerized scheduling programs. Once the schedule is established, you are also responsible for enforcing it, seeing to it that the work progresses smoothly and is done to the builder's satisfaction. You should expect to spend about 15 to 30 minutes per day updating the schedule. (Scheduling methods and techniques are discussed in more detail in chapter 5).

Staffing

Most superintendents have the authority to hire and fire in-house construction personnel and trade contractors. In addition, you may be responsible for training all site personnel and trade contractors regarding acceptable quality standards, operating procedures, construction methods, production efficiency, and company policies and procedures. Larger building firms may hire safety training managers to instruct employees on safety issues and OSHA compliance, while smaller builders delegate this responsibility to the superintendent.

Training employees and trade contractors is an ongoing process. If you use the same trade contractors on a consistent basis, your training responsibilities can be streamlined to some extent. If you use many different trade contractors or experience a period of high turnover in in-house personnel, you will need to devote more time and energy to training. In many ways your role as a trainer is like that of a teacher or an athletic coach; you must continually go back and cover basics as you bring your people along.

Evaluation

Industry experts in many businesses frequently say, "Superior managers get superior results." As a member of the management team, you receive part of the credit for building a profitable, growing business. Your performance as a superintendent should therefore be evaluated regularly according to previously established objectives. Figure 1.2 presents a self-evaluation form that will help you measure your progress and guide you in upgrading your performance.

If you find you are weak in some areas, you can work to improve your skills by reading, formal training, or education, and by consultation and conversation with other superintendents and builders.

FIGURE 1.2 Superintendent's Self-Evaluation

Rate yourself as to your strengths and weaknesses.	*Always*	*Sometimes*	*Occasionally*	*Never*
I schedule my own time and the time of others appropriately.	☐	☐	☐	☐
I am able to accomplish all of the important things each day.	☐	☐	☐	☐
I develop a schedule for each project before it begins.	☐	☐	☐	☐
I am aware of the status of all important aspects of the jobs.	☐	☐	☐	☐
I update construction schedules regularly.	☐	☐	☐	☐
I motivate others to do the designated amount of work in the designated time.	☐	☐	☐	☐
I complete projects on time.	☐	☐	☐	☐

Cost Control

	Always	Sometimes	Occasionally	Never
I establish a budget before construction begins on each individual project.	☐	☐	☐	☐
I am cost conscious.	☐	☐	☐	☐
I control costs discretely.	☐	☐	☐	☐
I foster cost consciousness in others.	☐	☐	☐	☐
I prepare material use or cut sheets.	☐	☐	☐	☐
I update the budget and calculate the cost variance regularly.	☐	☐	☐	☐
I identify the causes of cost variances properly and eliminate negative variances regularly.	☐	☐	☐	☐
I follow a systematic purchase order system.	☐	☐	☐	☐
I keep accurate records of all change orders.	☐	☐	☐	☐
I see that materials are properly stored in appropriate material storage areas on the jobsite.	☐	☐	☐	☐
I control waste.	☐	☐	☐	☐
I see that tools and equipment are properly taken care of.	☐	☐	☐	☐

Quality Control

	Always	Sometimes	Occasionally	Never
I have well-defined quality standards established for each project.	☐	☐	☐	☐
I communicate quality standards to each subcontractor, supplier, and worker.	☐	☐	☐	☐
I conduct systematic quality control inspections.	☐	☐	☐	☐
I am respectful of customers and responsive to their needs.	☐	☐	☐	☐
I control the quality of work performed on my projects.	☐	☐	☐	☐
I schedule and monitor the necessary inspections.	☐	☐	☐	☐

Safety

	Always	Sometimes	Occasionally	Never
I am safety conscious.	☐	☐	☐	☐
I display a knowledge of both company and governmental (OSHA) safety and health rules and regulations.	☐	☐	☐	☐
I follow the company safety program religiously.	☐	☐	☐	☐
I correct safety problems promptly.	☐	☐	☐	☐

(Continued)

FIGURE 1.2 (Continued)

Rate yourself as to your strengths and weaknesses.	Always	Sometimes	Occasionally	Never
I am aware of HazCom standards and ensure that they are followed on my jobs.	☐	☐	☐	☐
I keep my jobsites clean.	☐	☐	☐	☐

Organization

	Always	Sometimes	Occasionally	Never
I have an appearance of organization.	☐	☐	☐	☐
I get results by organization of paperwork.	☐	☐	☐	☐
My desk/truck is organized.	☐	☐	☐	☐

Delegation of Duties

	Always	Sometimes	Occasionally	Never
I delegate duties appropriately.	☐	☐	☐	☐
I assume responsibilities for those duties that I delegate.	☐	☐	☐	☐
I follow the lines of authority, as a supervisor and a subordinate.	☐	☐	☐	☐
I work well under pressure.	☐	☐	☐	☐
I maintain self-control.	☐	☐	☐	☐
I control the situation rather than permit the situation to control me.	☐	☐	☐	☐

Management Concepts

	Always	Sometimes	Occasionally	Never
I spend the necessary time planning.	☐	☐	☐	☐
I avoid management by crisis.	☐	☐	☐	☐
I set objectives.	☐	☐	☐	☐
I establish a plan to achieve the objectives.	☐	☐	☐	☐
I set major limitations and controls.	☐	☐	☐	☐
I measure my own performance as well as the performance of others on the basis of previously established objectives.	☐	☐	☐	☐
I delay decision making until I have investigated the answers.	☐	☐	☐	☐
I keep adequate daily records on each individual job.	☐	☐	☐	☐
I ensure that records are adequate enough to cover essential elements but not cumbersome or time consuming.	☐	☐	☐	☐
Trade contractors and suppliers enjoy working with me.	☐	☐	☐	☐
I coordinate all the essential elements of construction, including the performance of trade contractors, suppliers, and others.	☐	☐	☐	☐

Computers

	Always	Sometimes	Occasionally	Never
I am computer literate.	☐	☐	☐	☐
I use the following computer programs proficiently:				
word processing	☐	☐	☐	☐
spreadsheets	☐	☐	☐	☐
construction estimating	☐	☐	☐	☐

(Continued)

FIGURE 1.2 *(Continued)*

Rate yourself as to your strengths and weaknesses.	*Always*	*Sometimes*	*Occasionally*	*Never*
construction scheduling	☐	☐	☐	☐
cost control systems.	☐	☐	☐	☐

Interpersonal Performance

	Always	*Sometimes*	*Occasionally*	*Never*
I help establish performance goals for others.	☐	☐	☐	☐
I evaluate the performance of others on the basis of these goals.	☐	☐	☐	☐
I receive cooperation from others.	☐	☐	☐	☐
I am respected.	☐	☐	☐	☐

Training of Others

	Always	*Sometimes*	*Occasionally*	*Never*
I encourage training and establish the means for it.	☐	☐	☐	☐
I motivate others to perform to their maximum potential.	☐	☐	☐	☐
I have the courage and fortitude for hiring and firing.	☐	☐	☐	☐
I measure my own performance based upon the improvement of others.	☐	☐	☐	☐
My performance appraisal standards are objective.	☐	☐	☐	☐
I listen effectively, both to subordinates and to superiors.	☐	☐	☐	☐
I reward superior performance in the following ways:				
with praise	☐	☐	☐	☐
with consideration	☐	☐	☐	☐
with monetary means	☐	☐	☐	☐
with other types of remuneration.	☐	☐	☐	☐
I communicate or carry out requests of both subordinates and superiors.	☐	☐	☐	☐
I let others carry appropriate responsibility.	☐	☐	☐	☐
I step in and take over for others when it is necessary without adversely affecting their morale.	☐	☐	☐	☐

Personal

	Always	*Sometimes*	*Occasionally*	*Never*
I display common sense.	☐	☐	☐	☐
I have effective people skills.	☐	☐	☐	☐
I display a positive attitude.	☐	☐	☐	☐
I set goals for myself.	☐	☐	☐	☐
I seek self-improvement through seminars, study, schooling, consultations with others, or through other means.	☐	☐	☐	☐
I display the ability to become more of an asset to the organization.	☐	☐	☐	☐

2

Project Start-Up

As the company's field representative, your primary responsibility is to control and manage your projects in three areas: cost, schedule, and quality. You must effectively and conscientiously administer a construction program to meet the primary objective of maximizing profits and quality standards while you maintain positive business relationships with trade contractors, suppliers, inspectors, and—most importantly—homeowners. An effective, well-trained superintendent with the necessary authority and resources and a proper workload can exercise careful control, decrease costs, meet tight schedules, and ensure high quality work. Without the necessary authority, resources, and training, however, the inevitable result is loss of control, increased costs, prolonged schedules, and poor quality. The skilled superintendent uses the tools discussed here to maintain the necessary control.

Starting Off Right

A project that starts off right has a much better chance of finishing up successfully. On the other hand, a project that starts off on the wrong foot is almost always doomed to failure. Too often homes are built with little forethought and planning. Superintendents may be tempted to simply begin, assuming they will fit in the necessary planning as the work proceeds. Often a great deal of pressure is placed on superintendents to get jobs started in order to satisfy the homeowners or to meet production quotas and schedules. Homeowners have often been waiting for weeks from the time they signed a contract for plans to be prepared, estimates to be completed, and permits to be purchased. They are very anxious to see something "real" happen. Production departments often have production quotas and target schedules. If the superintendent gives in to such pressures, homes may be started before they are ready. Jobs that are started prematurely are likely to develop numerous problems when trade contractors fail to show up or arrive to find critical materials undelivered or the previous work unfinished. Frustration mounts, further complicating the job.

If you want a job to turn out right, you must spend the time planning for it up front. The key is to get ahead of the details and stay ahead as the project progresses. Far too often the job takes on a life of its own and ends up managing itself and the superintendent. The superintendent must then spend too much time, money, and energy putting out fires that should never have started in the first place.

Planning

To ensure a successful project start-up, do not cut corners with your preliminary planning and scheduling. Planning before starting construction frees your time later when the construction process demands much of your attention.

Construction Documents

To meet his or her responsibilities, every superintendent must be very familiar with the company's construction documents and with the details specific for each job, including the plans, specifications, and each subcontract agreement. You should have a complete understanding of the requirements of the contract and subcontract agreements, including:

- homeowners' (buyers') responsibilities
- scope of work
- price
- schedule requirements
- delivery of materials and supplies
- payment provisions
- lien laws

Preparing for the Job

One company that has received awards for being America's most organized builder and builder of the year insists on adequate preparation before construction begins. One of the conditions of employment for superintendents is to never start a job until the following tasks have been completed:

- the construction loan has been approved and closed
- plans are complete and have been approved by the homeowner(s)
- color selections are complete
- a site meeting has been held with the homeowner(s)
- a preconstruction meeting has been held with the owner(s)
- the estimate is complete and has been reviewed by the superintendent

- all purchase orders for the entire job have been sent to the various trade contractors and suppliers
- all permits and fees have been purchased
- the appropriate "call before you dig" authority has been notified and the property has been staked accordingly
- the property has been surveyed by a licensed surveyor

   If a superintendent jumps the gun or starts a home prematurely, he or she will be severely reprimanded, disciplined, or even fired. This may seem like a harsh penalty, but this company has found that, almost inevitably, if the above tasks are not completed, the project never gets back on track.

- liquidated damages, if any
- safety program documents and requirements
- termination of trade contractors
- change orders
- cleanup
- warranty provisions
- contract interpretation

Preconstruction

The superintendent acts as the catalyst for the preconstruction activities that bring the entire building process together. Other people are involved, but the superintendent should be the leader. Key elements in the successful planning of preconstruction activities include:

- plans
- specifications
- prequalification of buyer(s)
- site meeting
- customization of the home to the lot
- material and color selections
- plan review with homeowners
- preconstruction meeting with homeowners
- plan review by superintendent
- cost estimate
- identification of key personnel, trade contractors, and suppliers to be used on the project
- discussion of special requirements or administrative procedures that are unique to the job with key personnel, trade contractors, and suppliers
- plan review with key personnel, including review of the specifications
- review of initial estimate and purchase orders
- release of purchase orders
- possession of required permits
- establishment of start date
- hookup of temporary power

Site Meeting

One of the first things you must do before building is to make sure that the lot and the home fit together. If you are building a straight ranch-style home it is helpful to have a fairly flat lot. If you are building a home with a lower (basement-level) garage, it is nice to have a lot with a fair amount of slope. Flat lots and hillside lots require different planning. A lot in the country where a well and septic system are required has entirely different planning needs from a lot in a developed community.

As the company's jobsite supervisor, the superintendent is normally responsible for evaluating site considerations relating to the job. For example, the superintendent may need to evaluate conditions such as high water tables, unique drainage or excavation requirements, or areas where the lot's slope or soil conditions do not permit building a home according to the plan. The superintendent is usually the best resource for resolving on-site problems.

A site meeting is a valuable tool for resolving critical issues up front before construction ever begins. The superintendent should conduct the meeting at the site with the homeowners. Essential trade contractors, suppliers, and other people may attend as necessary, including representatives from the power company and health department (in rural areas where private wells and septic systems are required). Normally, the information from the site meeting is recorded on the plot plan. The following items should be considered during a site meeting:

- location of property pins
- location of the home on the lot, including setbacks and easements
- topsoil removal and replacement
- tree removal and clearing
- location and size of driveway
- location, size, and depth of utility trenches (water, electrical, gas, sewer, and cable television)
- location of power meter base and circuit breaker box
- location of temporary power and water
- establishment of existing grades and final grades in relationship to the house
- removal and stockpiling of topsoil as appropriate
- need for additional fill or existence of excess excavated material
- house drainage where applicable (sump system)
- subsoil conditions (water table and rock for homes with basements)
- location of well and septic system (in rural locations)
- location of air conditioner or heat pump
- most feasible method of trash disposal

Many builders conduct the homeowner site meeting before the preconstruction meeting. The superintendent meets with the homeowners on the site after the preliminary plans have been drawn and customized for the home. The superintendent conducts a complete site meeting and lays out the lot, locating the utility trenches and lines. While the homeowners are on the lot, the superintendent reviews the house plans in detail to make sure that everyone understands what the home is like and how it will fit on the lot. The superintendent is given a detailed plot plan of the lot, a cross-section of the house showing existing grade and projected final grade, and an excavation plan of the house. The superintendent reviews the plans, particularly the foundation plan, to locate the furnace, water heater, sump pump or natural drainage system, and other plumbing fixtures including the waste water pipe exit.

Project and Site Logistics

In addition to being responsible for and taking direct charge of site layout, superintendents also should be concerned with such site logistics as access, water, electrical lines, and other details. The site meeting checklist in Figure 2.1 will give you a model to use during a site visit.

Arrange for Utilities. Utilities must be readily available so that trade contractors and other workers do not have to waste time stringing long extension cords to obtain power or run long hoses to obtain water. Contact utility companies before the start of construction to arrange for temporary hookups at the jobsite. All electrical circuits in the temporary hookup should be protected with ground fault circuit interrupters (GFCIs). Temporary utility service at each jobsite will allow for specific job cost expense allocations.

FIGURE 2.1 Site Meeting Checklist

Job number: _____ Name: _____ Date: _____

Home phone: _____ Business/work phone: _____

Model of house: _____ Garage on: ☐ Left ☐ Right

Lot address (street and lot number): _____

Atlas map location: _____

City: _____ County: _____ Township: _____

Lot size: _____ Setback: _____ ft. From: _____

Sidelines seen from street: Left: _____ ft. Right: _____ ft.

PART I Explain and check either yes or no.

	Yes	No		Yes	No
Lot previously filled	☐	☐	Building specifications	☐	☐
Property pins located	☐	☐	Homeowners' responsibilities list	☐	☐
Lot surveyed by a licensed surveyor	☐	☐	Change order from site meeting	☐	☐
Owner pre-wire phone/TV	☐	☐	Reroute field tile	☐	☐
Easements/restrictive covenants	☐	☐	Zoning permit	☐	☐
Review contract	☐	☐	Dirt removed from drive	☐	☐
Discuss change orders	☐	☐	Additional excavation or cut swale	☐	☐
Discuss allowances	☐	☐	Drive/permit/culvert	☐	☐
Trash and cleanup	☐	☐	Clear lot, stumps, trees	☐	☐
Temporary downspout lines	☐	☐	No move-in until closing	☐	☐
Garage floor elevation	☐	☐	Recommend change location	☐	☐
Who and when owner should call	☐	☐	Electric for well pump	☐	☐
for information	☐	☐	Flood zone	☐	☐

Special comments: _____

_____

_____

_____

Homeowner(s) signature(s): 1. _____ Date: _____

2. _____ Date: _____

PART II Utilities

Electric company: _____ ☐ 100 amp ☐ 200 amp

Service location: _____ ☐ Overhead ☐ Underground

Application made: ☐ Yes ☐ No ☐ 100 amp ☐ 200 amp

Temporary pole: ☐ Yes ☐ No ☐ Overhead ☐ Underground

Panel location: _____

Mast pipe: ☐ Yes ☐ No Displaced service: ☐ Yes ☐ No

Underground service by: ☐ Homeowner ☐ Electric company

Sewer location: _____ Sewer depth: _____

Septic location: _____ Type of system: _____

Wastewater exit: _____ Through: ☐ Footing ☐ Foundation wall

Electric circuit needed for septic system: ☐ Yes ☐ No Supplied by: ☐ Homeowner ☐ W/H

City water: ☐ Yes ☐ No Well: ☐ Yes ☐ No

Location of water line into house: _____

Location of pressure tank: _____

Pressure regulator required: ☐ Yes ☐ No

Electric circuit needed for well: ☐ Yes ☐ No Supplied by: ☐ Homeowner ☐ Builder

Natural gas: ☐ Yes ☐ No Name of company: _____

Application made: ☐ Yes ☐ No

(Continued)

FIGURE 2.1 *(Continued)*

Length of exterior gas line: _____ Size of exterior gas line: _____

Liquid propane: ☐ Yes   ☐ No   Location of tank: _____

Length of interior gas line: _____ Size of interior gas line: _____

Location of gas line exit: _____

Footing sump: ☐ Yes   ☐ No   Laundry sump: ☐ Yes   ☐ No

Natural footing drain: ☐ Yes   ☐ No   Length: _____ ft.

Extra fill needed: ☐ Yes   ☐ No   Location: _____

Type of fill: _____   Supplied by: ☐ Homeowner   ☐ Builder

Extra block needed: ☐ Yes   ☐ No   ☐ Maybe   Amount: _____

Homeowner(s) signature(s):   1. _____   Date: _____

                             2. _____   Date: _____

PART III Superintendent

Indicate on the plans:

☐ Present grade
☐ Finish grade
☐ Extra block for foundation
☐ Footing sump location
☐ Laundry sump location
☐ Water line location
☐ Pressure tank location
☐ Waste water exit
☐ Electric service
☐ Electric panel
☐ Gas line exit
☐ Washer and dryer
☐ A/C or heat pump location

Indicate on plot plan:

☐ House location (set backs)
☐ Temporary pole
☐ All elevations
☐ Measurements
☐ Existing structures
☐ Propane tank
☐ All utilities
☐ Natural drain
☐ Drive
☐ Stake temporary power pole
☐ Post builder's sign
☐ Post MSDS sign

Special comments: _____

_____

_____

_____

_____

_____

_____

_____

_____

Superintendent's signature: _____   Date: _____

_____

Organize the Physical Layout. Each home should have a site plan indicating location of the home, temporary power, water, trash disposal, culinary water or well trench, sewer laterals or septic system, and so forth. Laying out the site on larger, custom-built homes involves the efficient placement of a storage area, drop sites for materials, perhaps a jobsite office, and areas for fabrication, if required. The objective of planning the physical layout is to ensure that work can be performed efficiently. Every physical site has its own special set of conditions that make this optimum layout unique. For example, your requirements in a cramped urban development or multifamily site will be much different from those in a rural location.

On scattered sites, a pickup truck and cellular phone may be the only "office" space available. In other situations a temporary field office and storage shed may be warranted. Your field office may be nothing more than a portable shed or designated area containing a small desk or plan table and a telephone. On a large project, a field office can be a particularly effective time-saver, giving the superintendent a central location from which to control the project. A storage area should be located where it will not interfere with other activities, yet still be as close as possible to the areas where the stored materials will be used.

Prepare for Surface Water Control. Once begun, grading should be completed as quickly as possible to minimize continuing disturbance of the soil. Open trenches should be covered as soon as possible. Make sure that straw bales or silt fences are ready for use in controlling and filtering runoff; rock or gravel also may be needed for filtering. Consider using temporary collecting ponds in the event of unforeseen, severe wet weather.

Preconstruction Meetings

One management tool that has been used almost universally in commercial construction and that is being adopted more frequently in residential construction is the preconstruction conference or meeting. The timing of the preconstruction meeting varies. Some builders hold the meeting

Jobsite Access

On rural lots, access to the jobsite is critical and sometimes very difficult. A temporary gravel driveway can ensure access to the site even during inclement weather. Often the temporary driveway becomes the foundation for the permanent driveway. If a temporary driveway is needed, its location and costs for creating and maintaining it should be discussed with the homeowner before construction begins. In winter months in many parts of the United States, snow-removal arrangements may be necessary.

immediately following the site meeting. Others wait a few days. A preconstruction meeting can significantly improve customer relationships and give homeowners the confidence that you are the right builder for them. The primary purpose of the preconstruction meeting is to set the expectation levels of all of the major participants in the project. A preconstruction meeting can do more to establish good working relationships, set homeowners' expectation levels, and prevent or resolve difficult problems up front than almost any other thing you can do.

The superintendent, homeowners, and perhaps the salesperson should attend the meeting. Because most residential builders use the same trade contractors and suppliers over and over again, most of these trade contractors and suppliers soon become familiar with the project policies and procedures. For a production home builder, therefore, it may be advisable to include only new trade contractors or suppliers in the preconstruction meeting. For a custom builder who may use different trade contractors for each job, however, it may be necessary to ask them all to attend the preconstruction meeting. On smaller jobs or for production housing, when a formal preconstruction meeting is deemed unnecessary, the superintendent should meet with all project participants individually to cover any necessary information. However the preconstruction meeting is primarily for the homeowners.

The meeting may be conducted by the superintendent, the production manager, or another member of the builder's staff, depending on company size and policies. The meeting site may be the construction office, sales office, company main office, or just about any convenient place where you will not be interrupted.

The meeting typically covers a company's routine policies and procedures for project management, written copies of which are distributed to the homeowners and each of the key project participants. A typical preconstruction meeting may last three to four hours or sometimes longer, depending on the complexity of the project and the needs and inquiries of the homeowners. Because one of the primary objectives of the meeting is to establish the responsibilities and expectations of each participant, lines of communication must be established and recognized by the participants, including both the regular channels of authority and communication in emergencies. Major discussion topics at a typical preconstruction meeting include the following:

- plans and specifications
- driveway location and makeup
- site plans
- permits and fees
- communication
- homeowner responsibilities
- allowances
- appliances
- HVAC systems and alternatives
- color selections
- insurance coverage
- materials and methods to be employed
- sequence and acceptance of work
- painting and decorating
- floor coverings
- grading and drainage
- utility hookup
- inspections
- construction schedules
- change orders
- trash disposal
- safety
- payment provisions
- bank draws
- disputes
- builder's and manufacturers' warranties and procedures

The homeowners, who were given a copy of the company's warranty agreement at the time of the purchase agreement, may be asked to review the warranty and sign it during the preconstruction meeting. The discussion may also include manufacturers' warranties on items such as mechanical equipment, roofing, siding, windows, and cabinets.

Planning the Preconstruction Meeting. Proper planning can ensure that this important meeting is more productive for everyone. Prepare an itemized agenda or checklist of items to be discussed and distribute it to participants at the beginning of the meeting so that they can reserve questions or comments until the appropriate time. Figure 2.2 provides an example of a detailed preconstruction meeting checklist. The meeting should be held in a convenient, comfortable place that is conducive to relaxed and focused discussion.

Conducting the Meeting. When conducting a preconstruction meeting you will arrange for or take the following steps:

- take minutes (yourself or someone present)
- begin on time
- establish ground rules
- define items to be discussed
- stick to the agenda
- maintain control of the meeting
- solve problems
- summarize decisions made at the meeting
- solicit questions and feedback from all participants
- set aside time for questions and answers

During the meeting, among other agenda items, you will examine and discuss the construction schedule. Be sure to point out critical dates for delivery of materials and completion of various phases of the project. Allow time to resolve any conflicts that may exist. After the meeting, distribute written minutes of the meeting to all participants.

FIGURE 2.2 Preconstruction Meeting Checklist

Review all items with each homeowner prior to the start of construction. If an item does not pertain to the homeowners, mark it "N/A" (not applicable). After the review is complete, have the homeowners sign and date pertinent documents. You may give a copy to the homeowners.

Items needed for review:

☐ **Buyer profile questionnaire:** Review to ensure that all necessary information has been provided.

☐ **Contract worksheets:** Review all contract worksheets to make sure the dollar amount is correct and that all items on the worksheet are also on the contract.

☐ **Sales contract:** Review the sales contract to make sure necessary information is filled in completely and correctly, and to review topics such as required insurance coverage, changes to the standard plans, and procedures for allowances.

☐ **Plans:** Go over the blueprint of the home in detail. Discuss dimensions; traffic patterns; location of appliances, light switches, and outlets; which areas will be finished or unfinished; and any other details that might cause confusion or might not be fully understood. Note any changes required on the plans. Make sure all of the changes to the standard plans have been incorporated into the final plans. Review and check or mark the location(s) of:
— Reverse plan (if applicable)
— Furnace and heat pump/air conditioner
— Heat registers and thermostat
— Water heater
— Pressure tank
— Electric panel
— Meter base
— Sewer line exit
— Well or water line entrance
— Cost, if option is purchased, for the following:
    $ _____ gas line
    $ _____ gas range
    $ _____ gas hot water heater
    $ _____ for gas dryer
    $ _____ for electric range
    $ _____ for electric hot water
— Purchase and installation of gas meter bar and gas line entrance
— Upgrades to standard furnaces
— Extra footage added to standard plans
— Stoop or porch size, overhang, columns, and so forth
— Basement wall height (note company standard)

— Basement plumbing
— Crawl space height
— Crawl space access door
— Brick, siding, stone
— Roof pitch
— Chimney (note applicable company standards)
— Fireplaces (note applicable company standards)
— Fireplace faces
— Insulative sheathing
— Door type, location, and swing
— Patio doors (active side)
— Entry sidelights
— Garage service door
— Garage door
— Window locations, extra windows
— Deck or ledger board size, location, construction, picket spacing, and so forth
— Hose bib number and location(s)
— Stairwell location
— Doors out of basement, lower-level (frost line)
— Insulation R-values
— Telephone and television jacks, locations and types
— Switches, including three-way switches, outlets, and ground fault circuit interrupters (GFCIs), if any
— Bath layouts
— Mirrors
— Plumbing fixtures (type and color)
— Tubs or showers (type and color)
— Cabinetry and vanities
— Countertops
— Range and refrigerator
— Drop-in range
— Built-in oven(s)
— Range hood
— Dishwasher
— Microwave(s)
— Kitchen cabinet outlets, types and locations
— Washer and dryer
— Floor covering types and location
— Material changes
— Other items: _____

☐ **Specifications:** Make sure specifications are included in the file. Review with the homeowners any questions about the methods and materials to be used in the construction process. Review financed construction draws and cash construction draws. Check for all required signatures and dates. Make sure that the homeowners have a set of specifications and that you have a signed copy.

(Continued)

FIGURE 2.2   *(Continued)*

☐ **Change orders:** Review the change orders already in the file. Make sure that all change orders, however minor, are completed and signed. Any changes to the contract must be authorized by signed change orders. Verbal changes will not be recognized. Explain the change order fee plus the price of the change. Have a change order form to fill out in case any changes result from decisions made at the pre-construction meeting.

☐ **Permits:** Discuss the acquisition of the necessary permits including the building permit and the payment of fees. Discuss the role of the homeowners in this process.

☐ **Insurance coverage:** Discuss builder's risk and homeowner's insurance and coverage during construction. Builder's risk insurance does not cover homeowner-furnished items or owners' personal items left onsite.

☐ **Model tour (if applicable):** Walk through the model and describe the construction process using the model as a visual aid. Try to make sure that the homeowners' concerns are addressed and that all questions are answered.

☐ **Owners' responsibilities:** Ask the homeowners if they have any questions concerning their responsibilities. Reassure them concerning their responsibilities but make sure they understand their importance. Make sure the homeowners' responsibility checklist is signed and included in the file.

☐ **Site meeting checklist:** Make sure the site meeting checklist is in the file and that all pages are properly filled out and signed. Explain plans for finish grade, drainage, and so forth. Write change orders as needed for any extra grading, fill, or other changes known at this time. Ask the homeowners if they have any questions as a result of the site meeting.

☐ **Utilities:** Discuss the location and availability of both temporary and permanent utilities and review the homeowners' responsibilities relating to applications or installation.

☐ **Site plan:** Briefly review the site plan. Make sure the home is properly located. Discuss potential problems inherent or unique to the site, such as drainage, location of septic system and utility lines, potential interference between construction operations and existing trees, and so forth. Review the homeowners' future plans for adding a pool, deck, or other features.

☐ **Color selection sheets:** Assist the homeowners in making their color selections and in selecting cabi-

nets, appliances, and so forth. Be sure any special items that may not be on the standard sheets also are covered. Make sure that the homeowners understand the importance of timely decisions regarding items such as lighting fixtures, paint, and wall coverings. Check to be sure color selection sheets are completed, signed, and dated by the homeowners.

☐ **Floor coverings:** Discuss in detail who will be responsible for which floor coverings and when they will be installed. If the builder will handle floor coverings, be sure the carpet selection sheet is completed and signed.

☐ **Construction schedule:** Review the construction schedule with the homeowners. Discuss the sequence of activities and the importance of the homeowners performing their responsibilities in a timely manner. Discuss painting specifically. Emphasize that your schedule will vary because of factors such as weather and inspections. To prevent confusion, avoid giving a written copy of the schedule to the homeowners. A basic breakdown of the major construction phases will include the following:

— Excavation

— Footings: Discuss digging and pouring footings and the time frame that is normal, whether or not there is a separate inspection for this task.

— Foundation (blocks or poured concrete): Note minimum curing time required for concrete walls before framing can begin.

— Framing: Discuss in detail and predict how long you think it will take to frame their house.

— Rough-in plumbing, mechanical, and electrical

— Exterior finish: Note that shingles go on before or during the plumbing rough-in and that the windows, exterior doors, vinyl siding, and brick (if specified) also is completed at this time.

— Interior finish: Explain that wall insulation and drywall are installed next. Be sure the homeowners understand the time required to put up drywall and the need for adequate ventilation, especially during the winter. Discuss interior trim and hardware installation.

— Painting: If the homeowners are doing their own painting, explain the process and the time frame they will have to complete the work.

— Cabinets: Review cabinet layout drawer locations, and so forth.

— Finish electrical and plumbing

— Floor coverings: Discuss the sequence of floor

(Continued)

FIGURE 2.2 *(Continued)*

covering installation. Review types of flooring in each area.

— Completion: Discuss the completion and punch out phase. Discuss the importance of one final punch list.

☐ **Communication:** Discuss the superintendent workload and review protocol for telephone calls (when and how). Discuss emergency communication procedures. Explain that the homeowners should not deal directly with the trade contractors at the jobsite but instead work directly with their superintendent to resolve any questions or situations that may occur.

☐ **Homeowners' orientation tour:** Briefly discuss the homeowners' walk-through and orientation.

☐ **Closing package:** Discuss the closing process and the homeowners' role in it. Note when keys will be given to the homeowners, discuss credits and charges that will appear on the closing sheet, monies to be paid at closing, and any other particulars.

☐ **Warranty and service:** Discuss the company's warranty service policies and procedures. Note the difference between warranties covered by the builder and warranties from product and equipment manufacturers. Discuss emergency telephone numbers and procedures.

Identifying Problems. Part of the purpose of the preconstruction meeting is to identify potential issues and problems. As problems come up during the course of the preconstruction meeting, keep your mind open to the alternatives and solutions suggested. Discussing problems without proposing or accepting viable solutions wastes time (unless further information is required). Methods of carrying out solutions, measuring results, and reporting back to the group must be established.

Schedules

The construction schedule will dictate the sequence of construction (which activities must follow one another). For example, if the interior trim is to be stained, it may be best to stain the trim and paint the interior walls before the trim is installed. However, if the trim is to be painted, then it would be more efficient to install the trim before painting.

Deliveries need to be scheduled so that materials are available when needed, but not so early that materials take up valuable storage space for an excessively long time—or disappear before they are used. Trade contractors should not have to move other contractors' materials to get to their own work or accommodate other contractors who may be on the site at the same time. Through careful planning, scheduling, and staging such conflicts can be avoided.

Regulations

Superintendents should always keep informed of local ordinances and regulations pertaining to their jobs and make certain to obtain all necessary permits. Discussing how to best comply with regulations, policies, and procedures with local building code officials will help you avoid problems at inspection time. Failure to comply with any regulation can mean considerable delays and, in some cases, fines for the builder. Inspectors may also require that perfectly good work be torn out and redone if the necessary requirements are not met.

Many superintendents keep a list of unique code requirements and changes by municipality. Different inspectors may interpret the same building code differently. Each may have his or her personal areas of concern. Keeping a list of particular inspectors' special concerns will help you to pass inspections the first time, avoiding unnecessary and costly delays and reinspections. Code officials' preferences are not necessarily code requirements. A thorough knowledge of the local code can help the superintendent in situations when an inspector requires questionable work.

Defining Trade Contractor Responsibilities

One of the superintendent's first responsibilities on a project is to assist in the estimating process by taking bids on work to be subcontracted and assisting in the selection of trade contractors.

Pre-Bid Procedures. Before obtaining trade contractor bids, develop your own, realistic ideas about how much each type of work should cost. Your own projections will assist in bid evaluation. When requesting bids, control trade contractor overlap and prevent gaps from occurring by clarifying and defining the exact scope of each subcontract or agreement. For example, does the footing trade contractor lay out the foundation on the footings, or is that job someone else's responsibility? Does the framer install felt paper on the roof, or is that job performed by the roofing trade contractor? Determine who is responsible for cleanup of trade contractor operations and removal of trash from the jobsite. These responsibilities must be clarified in order for construction to flow smoothly.

A well-though-out scope of work defining each trade's responsibilities helps facilitate communication and align the expectations of all parties. This sets the stage for positive, long-lasting relationships. The Home Builder Press publication *Contracts with the Trades: Scope of Work Models for Home Builders* contains model language and a further orientation to the use of written scopes of work that builders and superintendents can use with trade contractors (see Additional Resources).

Bid Review and Selection. Once you have developed a clear definition of each trade contractor's scope of work and an initial projection of costs, you can proceed with bid review, awarding of subcontracts, and major purchase orders. If so authorized, the superintendent should participate in evaluating all bids and awarding the contracts to the trade contractors and suppliers. (In some companies, this part of the process may be done by the estimators or reserved exclusively for the builder.) At minimum, if you will be responsible for trade contractor coordination you should act within your authority to ensure that each trade contractor understands the lines of authority and your position of responsibility. Keep in mind that the lowest bid is not always the best bid.

Other Responsibilities. As the field superintendent you serve as a resource for the trade contractors, answering their questions, clarifying instructions, and solving problems. In addition, you are responsible for ensuring that all necessary equipment is safe, adequately maintained, and available to those who need to use it, including necessary accessories such as oil, saw blades, and power cords.

Construction

A superintendent's administrative responsibilities are just beginning when construction begins. A well-implemented system of documentation can make job control easier during construction and leave a valuable paper trail for each project.

Reports and Documents

Superintendents sometimes attempt to control a project with little more than a vague recollection of past performance on other projects. "Seat of the pants" management is a major cause of poor control on a construction project, as are poor reporting practices. On the other hand, superintendents who find themselves continually deluged with paperwork may lose perspective, motivation, and control over their projects. A balance between these two extremes allows the superintendent to accurately build on past experience and focus on appropriate details.

Effective design and use of documents and reports can provide the needed balance. Reports that document what actually occurs on a project can function in the following ways:

- inform company management of project status at any given time
- establish projections for future activities
- assist in job control
- reduce the likelihood of a problem recurring
- give the superintendent a paper trail to resolve disputes
- become critical legal evidence

Adequate documentation also is necessary in managing change orders. Without proper documentation, pricing change orders and requesting payment for items beyond the scope of the contract quickly becomes impossible.

Accuracy and Completeness. For construction documents and reports to have any legal credibility, they must be accurate. Be careful to note any specific details immediately to ensure the accuracy of all documented information. Also make certain that reports contain all information needed to properly manage and control the job. Remembering everything that happens every day on multiple construction jobsites is an impossible task; however, you should document in writing any important occurrences at the time they occur. Some superintendents carry a pad of paper, notebook, or day timer and are continually jotting down items they need to remember. Others carry a small tape recorder, dictating as they go to eliminate confusion and the time involved in writing. (Labeling and maintaining the tapes, however, can be a real hassle.) Still others use small notebook computers to keep records and information. Notebook computers are now relatively inexpensive and can save you a lot of time. Storage of information is not a problem and reports can be easily printed.

Objectivity. Reports must be objective to be acceptable. Facts should be presented without interjecting opinions or being defensive. The credibility of a blatantly biased report may be called into question if a dispute arises.

Uniformity. Reports should be provided on a standardized form or in a standardized format, facilitating comparisons with previous reports.

Believability. By documenting all necessary items in a timely and unbiased manner you establish credibility as a superintendent. Most people can readily discern an altered or biased report. Handwritten reports are generally more acceptable in legal proceedings when they are documented and bound in such a way that pages cannot be moved or added. Original notes should not be modified; if something is forgotten and remembered later, add this information to another report at a later date rather than altering the earlier report. If you must amend an original report, draw a line through the original language rather than erasing it, and date your correction.

Timeliness. Timely reports are one of the key elements to maintaining or improving communications and eliminating misunderstandings. For management to make informed, effective decisions, superintendents' reports must be up-to-date and submitted regularly.

Types of Reports

Carefully maintained reports documenting daily activities and long-term project status can be important tools for the superintendent, the builder, and the home-owners.

Daily Reports. The daily report or daily jobsite activity log is generally a hand-written account, preferably kept in a bound book (see Figures 2.3 and 2.4 for examples of content and format). Daily reports have greater legal credibility than nearly all other accounts of job activity. Therefore, you must maintain these reports accurately, taking care to include all items of importance and particularly any item that might become the subject of a disagreement.

Daily reports have six basic purposes:

- to provide a record of activities including trade contractor performance, work completed, inspections completed, and so forth
- to make immediate note of instructions given verbally to ensure that action is taken and that the action is consistent with the instructions
- to document any verbal commitments that are made
- to back up future change order requests and additional charges
- to provide a well-kept written record for possible use in the event of a dispute settlement, arbitration, or lawsuit
- to document the site conditions and weather along with any delays or disputes

Progress Reports. These reports summarize the status of each phase of the project (see Figure 2.5). Progress reports perform the following functions for project managers and builders:

- to communicate project status to company management and others
- to summarize information contained in daily logs
- to outline instructions and decisions made regarding trade contractors and suppliers
- to summarize progress compared with the schedule
- to allow coordination between the office and project superintendent by assembling necessary information in an organized fashion

The progress report and the daily report should complement each other, with the progress report summarizing the daily report and communicating necessary information to various parties involved in the project. Progress reports may be sent to trade contractors, suppliers, the homeowners, and the design professional.

FIGURE 2.3 Items to include in the Daily Report

Items need not be limited to this list.

Job Conditions

- A complete description of the day's weather, including high and low temperatures, average temperature, and wind speed, in as much detail as necessary to document conditions

- Notations describing any problems that arise related to utilities, access to the site, drainage, snow removal, wet or muddy conditions, and anything related to site conditions and facilities

Visitors

- A log of all visitors to the jobsite who are not directly involved in the construction process, such as inspectors, homeowner(s), architects, building officials, safety inspectors, or government officials

Activities

- Information received from the owner or design professional, including the selection of items, colors, change orders, and accessories

- Work started, completed, or in progress

- Quantity of work performed (for example, cubic yards of concrete placed, amount of bricks installed) to assist in job cost control

- Labor problems, including specific disputes or disagreements

- All accidents (no matter how minor), including who was involved, witnesses, the circumstances surrounding the accident, the results, and any implications for safety authorities

- Major material deliveries and or delays

- Purchase orders issued from the field or material purchased

Trade Contractors

- Trade contractors who start or complete work, and contractors on site with work in progress

- The number of employees each trade contractor has present on the jobsite

- Questions to and from trade contractors

- Instructions given to trade contractors (follow up with memorandums)

- Any disagreements with trade contractors or suppliers on the jobsite

Communications

- All change orders requested and the status of each change

- Directives given by owners, inspectors, the builder, or architect

- Any unusual circumstances or problems encountered as a result of these directives

Schedule

- Work completed

- Work in progress

- Work to be started in the immediate future

- Problems relating to the schedule

Equipment

- All equipment present on the jobsite that day

- Equipment needs

- Rentals and returns

- Breakdown and repairs

- Other equipment problems

Contacts

- All phone conversations or personal contacts, the circumstances of each conversation, and the results, including any actions to be taken as a result of the conversation

FIGURE 2.4 Sample Daily Report Form

Project: _____   Weather: ☐ Fair ☐ Overcast ☐ Rain ☐ Snow
Job number: _____   Temperature: ☐ 0-50° ☐ 50-80° ☐ 80+°
Superintendent: _____   Wind: ☐ Still ☐ Moderate ☐ High
Workforce (trade contractors):
☐ Foreman   ☐ Plumbers   ☐ Bricklayers   ☐ Cement finishers
☐ Foundation   ☐ Electricians   ☐ Roofers   ☐ Floor covering installers
☐ Framers   ☐ HVAC   ☐ Tile installers   ☐ (Other)
☐ Carpenters   ☐ Insulators   ☐ Painters   ☐ (Other)
☐ Laborers   ☐ Drywallers   ☐ Cabinetmakers   ☐ (Other)
Equipment on job: _____

_____

Remarks: _____

_____

Visitors
Time   Name   Representing   Remarks

_____   _____   _____   _____

_____   _____   _____   _____

_____   _____   _____   _____

_____   _____   _____   _____

Equipment needs, rentals, problems: _____

_____

Work completed (describe): _____

_____

Work in progress (describe): _____

_____

Remarks, telephone conversations, contacts, problems: _____

_____

Work in progress (status): _____

Superintendent's signature: _____   Date: _____

FIGURE 2.5 Items to Include in a Progress Report

Communication to the Homeowner

- Decisions and actions required from the owner
- Delays experienced and their causes
- Verbal instructions and information received from the owner
- Change order requests and their status

Information for Trade Contractors and Suppliers

- Status of each contract or major purchase order
- Summary of instructions and changes that occurred during the week
- Items requiring coordination in current work

- Notice of future performance requirements or schedule changes

Schedule Progress

- Summary of current status of the schedule (as nearly as possible)
- Comparison of the current status with the scheduled status of the project
- Outline of any variances
- Description of the impact of homeowner-caused delays on the project

Problems

- Summary of problems, their apparent causes, and proposed solutions

3

Quality Control and Inspections

"**S**hoddy housing construction is a national plague. . . . The problems are pervasive, and shoddiness is a pattern, not just an occasional lapse of workmanship. . . . A serious problem exists, but it is a result of a minority of builders." These words, spoken by a leading consumer activist, underscore the importance of quality control for all builders. In the face of negative publicity, honest, well-run building companies face an uphill battle to gain and maintain homeowner trust and loyalty. The superintendent has a critical role to play in quality control and inspections. Superintendents' quality control efforts create and sustain the solid underpinning on which the builder's positive reputation must rest. Your quality control efforts help the builder overcome negative stereotypes about the building industry, win over homeowners, and create loyal homeowners.

Why Is Quality Control Necessary?

Serious or recurrent quality control problems in residential construction often can be traced to one or more of the following causes:

- incomplete contract documents
- inadequate specifications
- inadequate supervision
- inadequate inspections
- underqualified workers
- poor attitudes
- acceptance of inferior work

Incomplete Contract Documents. Too often, the plans and contract documents prepared for building a house are totally inadequate. They lack details and critical information. Plans may be drawn by draftspeople who lack experience in construction. Often they are based on pictures from a magazine and do not reflect the type of construction in the area of the country where the home is ultimately built.

Inadequate Specifications. Seldom are homes built using detailed written specifications. A frequent omission is material descriptions. Homeowners often do not know whether their doors will be masonite, birch veneer, or six-panel oak. They don't know whether the doors will have two hinges or three, or whether the closet doors will have wooden jambs or be wrapped with drywall. When descriptions are included in specifications, they often contain phrases like, "in a workmanlike manner" or " the best quality available." These phrases really do little to clarify acceptable construction standards.

Inadequate Supervision. Foremen, trade contractors, and superintendents need to be adequately trained. Superintendents need to be trained in all areas of construction, including plumbing, electrical, and heating, ventilation, and air conditioning (HVAC).

Inadequate Inspections. Superintendents are busy people. It is impossible for them to be on every jobsite all of the time. They may not see a potential problem until it is covered up by later work. Inspections by the superintendent are not the answer. Each worker must assume the responsibility for his or her own work. Trade contractors must each inspect their own work before the next contractor comes in. Inspections by the superintendent remain necessary, but are secondary to the personal inspection of the worker and his or her supervisor or employer.

Underqualified Workers. In a recent survey builders across the nation were asked to identify their number one problem. The biggest problem identified in the survey was a shortage of trained workers. The current labor shortage has been forecast to continue for at least ten years and is projected to become worse until well into the next century. There are simply not enough workers to fill the positions. Further, the home building industry is not training workers to fill the positions available. Many apprenticeship programs in the home building industry have been discontinued. As the industry switched from in-house labor to subcontracted labor in the late 1970s and 1980s, the trade contractors did not initiate training programs. As a result, most training is now done on the job, by trial and error. Many of the homes we build are filled with the training mistakes of new workers.

Poor Attitudes. Poor attitudes are a major cause of problems in quality. Often, however, the attitude problem starts at the top of the company. Many high level managers see the primary objective as getting the homes built on schedule. This is an important goal, but it cannot come at the expense of quality. Work must be done right the first time. It is too expensive to come back and do it over again. The reputation of the builder is the key to growth and profitability. We must foster—and reward—an attitude that promotes high quality work in all of the workers who build our homes.

Acceptance of Inferior Work. In order to make a profit, some builders and managers resign themselves to accepting inferior work. Building homes is a tough business. The average profitability of home builders nationwide measures less than 5 percent. Small contractors often enter the business with relatively little experience, little overhead, and even less understanding of the importance of accurately accounting for all costs. These small businesses have a tough time competing with larger builders who have more resources and receive volume discounts. Consequently, profit margins are low.

Many builders find themselves cutting corners to make up for costs they failed to consider when bidding a job. When this happens, inferior quality or slipshod construction sometimes results—not from malice, but as a matter of survival. To reduce this problem the industry must train builders to more accurately estimate the cost of construction and better record material, labor, and overhead costs. Quality should never be sacrificed in order to maintain profits. Indeed, high quality typically pays—it does not cost.

Callbacks to re-do work already completed are costly for any construction company. Recent statistics published in NAHB's *Nation's Building News* indicate that the average number of current problems per completed house has risen to four. For about ten years the average had been just three, so the number of callbacks has recently gone up by fully 25 percent. More than a third of the larger builders are reporting at least six problems per house. Most of the problems are minor, builders say. But the average cost to the builder is now $332—up 7 percent from $309 in 1997 and a whopping 17 percent from $283 in 1996. Among many of the larger builders, the average cost per callback has risen to $500 or more. Furthermore, the average response time has reached nine days, up from eight days in 1997 and six days in 1996. And it now takes more than three weeks to react to some calls.

Four callbacks costing an average of $332 per callback adds up to a total of $1,328 per house. If you multiply that by the number of houses built per year, the number begins to be alarming. Audits of some companies show warranty costs sometimes run as high as 2 percent or 3 percent of sales. That's a lot of potential profit going into warranty work. Indeed, that is half of the profit in an average company! (Many companies' warranty costs are 1 percent or less of sales.) But one of the biggest—and least easily quantified—costs of high amounts of warranty work is the bad reputation it can give to a company, especially if corrections are not made in a timely manner. A reputation for building high quality homes that brings homeowner referrals is of immeasurable worth. Likewise, a bad reputation can do an immeasurable amount of damage to future business.

The Superintendent's Responsibility for Quality

Of the superintendent's three primary responsibilities, quality control has the greatest impact on the overall long-term success of the company. The builder's reputation and future success will depend greatly upon the quality of work performed. Most builders depend heavily on referrals for future business and referrals come from high quality work. Producing a high quality product is therefore one of the primary responsibilities of management. Blaming material suppliers, craftsmen, and trade contractors for a lack of quality is a cop-out. The superintendent is the one who sets the standards of performance. Inadequate material should be rejected. Craftsmen and trade contractors will typically perform the quality of work expected and required by the superintendent. The superintendent must be able to differentiate between high quality work and inferior work. In the long run, quality of craftsmanship and the ability to perform within the available time limits may be far more important than price.

The manager who does not know how to recognize high quality in plumbing, electrical, heating and air conditioning, concrete, or finish carpentry is really not qualified to be a superintendent. A superintendent must have at least a basic understanding of each of the trades and the quality required. He or she must know the basic codes relating to the subcontracted work and require that trade contractors produce high quality work. The axiom, "the minimum you expect is the maximum you will get" often proves true.

Quality control on the construction site rests on the superintendent's execution of the following basic tasks:

- establishing required performance standards for both materials and workmanship
- establishing procedures to ensure that these standards are met and verified
- making sure that completed work meets the established standards

The best superintendents learn to judge high quality work through years of valuable experience. However, less-experienced superintendents can upgrade their skills by taking the following steps:

- talking with experienced superintendents and asking questions
- reading books and articles about quality management and about specific trades
- using educational audiotapes and videotapes
- talking with municipal inspectors
- seeking further education through industry or higher education channels

Total Quality Management (TQM)

A wealth of information on quality is available. The excitement surrounding the Total Quality Management (TQM) movement has inspired numerous books and publications, some written for our industry and published by NAHB (see Additional Resources). *Professional Builder, Builder,* and other trade publications also frequently feature excellent articles on quality management.

The TQM movement has had a tremendous impact on the quality of construction in the past decade. It has focused the nation's attention on the need for measuring performance in order to improve quality. W. Edwards Deming, the father of TQM, proposed fourteen basic tenets which, if followed, will result in significant improvements in quality:

1. Create constancy of purpose toward improvement of product and service, with the aim to be competitive, to stay in business, and to provide jobs. Create and publish to all employees a statement of the aims and purposes of the company or other organization. The management must constantly demonstrate their commitment to this statement.
2. Adopt a new philosophy. Top management and everybody. Western management must awaken to the challenge of a new economic age, learn their responsibilities, and take on leadership for change. We need a new "economic religion" in which mistakes and negativism are unacceptable.
3. Cease dependence on inspections to achieve high quality. Eliminate the need for mass inspection. In effect, American business has been paying workers to make defects and then correct them. Build high quality into the product initially. Understand that high quality comes not from inspection, but from improvement of processes and cost reduction. Workers need to be trained to be a part of process improvement.
4. End the practice of awarding business on the basis of price alone. Instead, minimize total cost. Rework is costly. Move toward a single supplier (trade contractor) for each item or trade, based on long-term relationships of loyalty, trust and high quality workmanship.

5. Improve constantly and forever the system of production and service. Quality improvement is not a one-time effort. Improvement of quality is a continual process of measuring how you are doing and finding methods of improving performance. Management must constantly look for ways to reduce waste and improve overall quality.

6. Institute job training. Too often workers learn their job from another worker who was not thoroughly or properly trained either. Train all employees. Develop skills in new hires and assist management to understand all processes of the organization.

7. Teach and institute leadership. The job of management is not to tell people what to do nor to punish those who don't perform. The job of management is to lead! Supervision of management and production workers should help people (trade contractors) work together to do a better job.

8. Drive out fear. Many employees (trade contractors) are afraid to ask questions or take a position even when they do not understand what the job is or what is expected. Create trust. Create a climate of innovation.

9. Break down barriers between departments. In construction there is often a rift between sales and production, or between estimating or drafting and production. There seems to be much more competition than collaboration. Optimize the aims and purposes of the company, the efforts of teams, groups, and staff departments or areas. Teach people to work together as a team. Measure the results of teams as well as those achieved by individuals and departments.

10. Eliminate targets, exhortations, and slogans for the workforce. Slogans alone never helped anyone do a better job.

11. Eliminate numerical goals and quotas for production. Instead, learn and institute methods of improvement. Learn what processes accomplish and how to improve them.

12. Remove barriers to pride of workmanship that rob hourly workers, as well as managers, of their right to provide high quality workmanship. Remove faulty equipment and methods that hamper employees and stifle innovation. Eliminate the annual rating or merit system.

13. Institute a vigorous program of education and self-improvement for everyone. Both management and the workforce, including trade contractors, will have to be educated and trained in this new paradigm of continual improvement.

14. Institute an action plan that puts everyone in the company to work to accomplish the transformation. Success will require commitment from top management as well as the rest of the organization: Workers can't do it on their own, nor can managers. You need each other. Assess where you are now and institute a program of continuous improvement companywide. For continuous improvement you will need to involve trade contractors in the process.

Obstacles to Total Quality

Deming also identified what he called "deadly diseases and obstacles" to total quality, which include the following:

- lack of constancy of purpose
- emphasis on short-term profits, which undermines quality and improvement and leads to neglect of long-range planning
- evaluation by performance, merit rating, or annual performance review

- mobility of management
- reliance on technology to solve problems
- managerial focus on visible numbers alone
- passive dependency on models or examples rather than innovative development of solutions
- excuses such as, "our problems are different," or "we have tried that before"

Some of these obstacles are obvious; others merit a word or two of explanation: performance rating systems may seem attractive in theory, but in practice they can destroy teamwork, nurture rivalry, build fear, and leave people bitter, despondent, and beaten. Some managerial mobility can be positive—but "job-hopping" managers do little to improve the organization. They are often not in a position long enough to understand the organization, the process, or the problems, which is necessary to bring about the desired improvement. Long-term changes are necessary to sustain improvements in quality and productivity. As for running a company by the "visible numbers," often the most important numbers are unknown and unknowable—for example, the multiplier effect of a happy homeowner.

As builders across the nation have begun to implement the concepts promoted by Deming many have had tremendous success. Most acknowledge it is not an easy process, and for some it has been a downright difficult process. People's attitudes and habits do not change overnight. But most builders have found that focusing on quality improvement has brought about significant changes in the organization and a tremendous increase in homeowner satisfaction.

Remember that quality control is not the superintendent's exclusive domain; it requires the participation of everyone involved in the construction process: architects, homeowners, contractors, salespeople, superintendents, suppliers, inspectors, lending institutions, and especially in-house workers and trade contractors. All members of the building team have a responsibility to ensure that their work meets the mutually accepted standards and that the materials used are of a quality equal to—or better than—the materials specified. All members of the team should continually look for ways to improve overall quality and not be content with meeting today's standards.

Creating an Atmosphere of High Quality

The first step superintendents take to ensure a high quality job is to establish required performance standards for both materials and workmanship. These standards must be communicated internally and also communicated to homeowners to establish their levels of expectation. Larger builders use model homes to demonstrate their quality, style, and craftsmanship. Smaller builders often use mockups, photos, production manuals, and other means to demonstrate what they will produce for the homeowners.

Attitudes directly affect the quality of construction work. To achieve a high quality operation the builder must have a deep personal commitment to serving homeowners and must foster pride in the kind of work that is performed. Homeowners have the right to quality in construction, and quality need not—should not—be sacrificed to profit. As the builder's right arm, the superintendent must share in and promote this commitment to quality.

Attitudes start at the top of any organization. The emphasis on quality control should begin with the president or CEO and trickle down through the entire company structure. Almost every worker possesses an innate desire to produce high quality work, and superintendents should make every effort to cultivate this attitude and motivate the worker. A company that emphasizes high quality inspires excellent workmanship. Without the active support of upper management, however, mottos, slogans, and signs mean nothing—they are simply wasted effort.

If trade contractors are selected on their ability to perform quality work, not price alone, the quality of their work will raise the level of the quality of everyone else's work. If superintendents consistently train trade contractors to perform to a specific standard of work, consistently monitor the quality of work performed, and make sure that all work is checked and completed before payment is made, quality overall will improve.

> ### PRIDE
>
> **O**ne company has implemented a program called PRIDE: Personal Responsibility in Daily Excellence. PRIDE is aimed at motivating individual workers to peak performance. The program lets employees know how important they are and helps them develop new confidence and self-esteem. By taking responsibility for their own work, employees satisfy their own needs for recognition. In addition, employees responsible for improved quality are given sizable monetary rewards.

You may have heard a modern version of the golden rule, "he who has the gold rules." In this case the builder has the gold! The superintendent makes sure trade contractors perform according to established standards before payment is made. A small item in chapter 2 highlighted one company's "conditions of employment" for superintendents. One condition is to never start a job before it was ready; the other is to never pay a trade contractor before the work has been properly completed and adequately inspected. In that company if a superintendent approves payment to a trade contractor without inspecting the work it could cost the superintendent his or her job. When you get that serious about quality, wonderful things begin to happen.

Performance Standards

Written performance standards are the basis of any quality control system. These standards communicate to those involved what is to be built, the criteria to be used in judging quality, and the specifications required. Management should devote the time necessary to make certain that all standards are precise, understandable, and based on measurable criteria.

Every company should establish standards of quality for each trade. Many companies find it useful to outline these standards in a manual. The National Association of Home Builders (NAHB) has developed a *Production Manual Template,* available in hard copy or on computer disk, that outlines the construction standards for each trade or phase of construction (see Additional Resources). The manual is full of photos and diagrams that depict both good and poor practices. It also contains a detailed quality control checklist for most of the trades. Because the manual is designed as a template, builders can use the information to develop their own production manuals, custom designed to the type of work they perform.

In the long run, the only way to achieve high quality is for each individual who works on the home to do his or her job and do it right the first time. Of course, the various trade contractors are responsible for the performance of each of their employees. The superintendent is in turn responsible for each of the trade contractors. With this kind of approach, quality control checklists can be an important tool that fosters self-monitoring for higher quality (see Appendix).

Quality control checklists are tools for everyone to use in monitoring the quality of the work. They are not all-inclusive and are not intended to be the final answer in determining quality. They present minimum standards to be met or exceeded. A quality control checklist can never replace common sense and pride in workmanship. The trade contractor or his or her employee should know what constitutes quality construction for that particular trade. The superintendent should be able to recognize quality in all areas.

A practical and successful quality control program incorporates the following elements:

- The trade contractors and each of their employees commit themselves to producing a quality product from the start.
- Each trade contractor inspects the work of its own crew on every job before they leave the job to verify that the work has been properly completed, represents high quality work, complies with all applicable building codes, zoning ordinances, health department regulations, and any other applicable regulatory agency requirements, meets or exceeds the basic standards in the subcontract agreement, and meets the minimum standards of the builder's quality control checklists.
- As applicable, the trade contractor completes the quality control checklist, reviewing all or part of the information with the superintendent.
- In some instances (for example, framing and interior trim) the superintendent approves the work before the trade contractor can leave the job.
- The trade contractor submits the quality control checklists to the superintendent upon completion of the work, as applicable.
- The superintendent inspects the work of the trade contractors daily as he or she makes his or her rounds.
- The superintendent turns in the completed quality control checklists to the production manager or supervisor at the weekly production meeting.

Training

A builder's established quality standards must be communicated both to the homeowners and to the workers responsible for building the home. The superintendent should provide each employee and trade contractor a copy of the standards and any applicable quality control checklists. New trade contractors should be trained before they start work on their first job for the builder. The superintendent should explain the quality standards and work with the new trade contractor to make sure the contractor and his or her crew understands them. Many superintendents have found that they must spend 50 percent of their time with a new trade contractor on the first job they produce. If a new framing contractor is framing a house for your company for the first time, for several days you may have to spend half of your time working with the framers. Presumably they are skilled at their craft, but they don't yet know your company's materials and methods.

During initial orientation and training of new contractors, superintendents should make themselves available to answer questions. Doing so helps make sure the trade contractors have confidence and trust in you.

Continued training also is essential. Many trades experience a constant turnover of personnel. For example, how often do you see the same person installing insulation job after job? Trade contractors' new employees need to be trained. Of course, training of the sub's employees is primarily the responsibility of the trade contractor— but the superintendent must make sure it happens. Consider taking a creative approach to quality management. Host an annual or semi-annual breakfast meeting with trade contractors to re-emphasize standards and mutual benefits of quality. Call it a "trade contractor appreciation breakfast."

Training of in-house employees also is necessary for large and small building firms. Superintendents should have a defined training program for each employee and a means of determining the effectiveness of your training. Record when the individual was trained in each area. Training of in-house employees begins the day they are hired. It continues on their first job and through continuing group training sessions and informal discussions, or through seminars. Superintendents can also use informal inspections and critique sessions to coach new employees.

Measuring Performance

Thomas S. Monson, a businessman and religious leader, once said, "When performance is measured, performance improves." The mere measurement of performance will invariably result in improvement. This phenomenon has been proven over and over again in experiments and in industry.

Monson also said, "If performance is measured and feedback is given, the rate of improvement will accelerate." In other words, if you tell people how they are doing, not only will they improve, they will improve at a faster pace. This phenomenon certainly applies to production. In the parlance of TQM, some people call it *benchmarking*. One of the major problems in construction is that the people performing the work really don't know how they are doing. Trade contractors especially are kept in the dark. They may assume that no news is good news. They seldom receive feedback unless they make a mistake.

Builders keep performance records in many different areas. The superintendent's daily log and progress reports are important tools in maintaining adequate records, particularly of recurring problems. The builder needs to know, for example how many failed building code inspections occur and why each one didn't pass. Another important area to track is how many callbacks the company experiences, preferably logged by superintendent, by trade contractor, by frequency of the problem, and by cost (see Figure 3.1). Homeowners should be surveyed considering their satisfaction with the builders, product and process.

What are your company's most frequent problem areas? Which are most costly? Which trades are in need of training? Which trade contractors seem to have their quality act together and can be resources to others? The information can be graphed to highlight trends.

FIGURE 3.1 Most Common Callback Items

Work	Builders (percent)	Work	Builders (percent)
Drywall	51	Heating, air conditioning	12
Paint, caulking	44	Foundation, basement	9
Roof	25	Electrical	5
Plumbing	23	Appliances	4
Doors, windows	19	Exterior walls	3
Grading	18	Driveway	2
Floors, walls, ceilings	12	Septic, sewer system	1

Source: Survey results in an article by Lew Sichelman, *Nation's Building News,* July 20, 1998.

Many builders find it helpful to use customer surveys to keep tabs on homeowner preferences and to identify areas in which the builder can improve customer service during the construction or warranty periods. A sample customer survey appears in Figure 3.2.

Internal Inspections

Builders who lack formal quality control systems rely heavily on public inspectors to detect faults, provide punch lists, or identify corrections. This practice is risky and unnecessary. Each company should conduct its own internal inspections if it wants to reduce the risk of inferior work and expensive rework.

If all work is inspected, corrected as needed, and approved as the project progresses, fewer problems will remain or arise at final completion. Internal inspections must be timely if they are to be of any value. As the superintendent responsible for quality control, you should actively inspect each project daily, even if projects are on widely scattered lots. Verify whether the work is progressing as scheduled and check work completed since the last visit.

Correct inferior work immediately. If the homeowners point something out to you, fix it immediately. If you don't, the homeowners will focus on the mistake each time they visit the job and will begin to wonder what else they may have missed. If you postpone fixing such problems until it is more economical to correct them or try to bunch them all together and fix them all at once, the homeowners will be much more dissatisfied and less trusting. If your homeowners begin to keep a list of concerns, you know you're in trouble.

Homeowners generally don't expect everything to be perfect. Normally they are reasonable people. If you listen to their concerns, give proper attention to the issues raised, take steps promptly to correct legitimate problems, you will impress the homeowners with your attention to detail and to their concerns. As a result, the homeowners will likely be less picky and more satisfied overall. Their trust in you and the company will increase, and the final walk-through will go much more smoothly.

When handling callback items have the person who caused the problem correct it. This practice makes those who perform the work accountable for quality. If a third party, such as a punch-out person, always corrects the mistakes, your trade contractors and their employees can become complacent. They will know they need not take your quality standards seriously because they know you will follow behind them and clean up their messes. On the other hand, if you cannot get a trade contractor to correct the problem quickly, it is better to correct it yourself rather than risk losing the confidence of your homeowners.

Inspection Points

Detailed internal inspections are very important, particularly at the following points during construction:

- before placing concrete footings and foundations
- before placing concrete floors
- on the final day of rough framing (while the framer is still finishing up)
- upon completion of rough mechanical work
- after rough electrical and plumbing
- when wall insulation is being installed

FIGURE 3.2 Sample Customer Survey

Prestige Home Builders   Survey # 81

Customer Satisfaction Survey

Thank you for taking the time to complete this questionnaire as part of our continuing desire to evaluate and improve our services.

How would you rate the information and service you received...	Excellent	Very Good	Good	Fair	Poor	
Before Purchase	⊙	⊙	⊙	⊙	⊙	☐
During Construction	⊙	⊙	⊙	⊙	⊙	☐
During Closing Process	⊙	⊙	⊙	⊙	⊙	☐

Please mark the ONE MOST IMPORTANT item affecting your satisfaction ──

How would you rate your SALESPERSON on each of the following:	Excellent	Very Good	Good	Fair	Poor	
Concern for your needs	⊙	⊙	⊙	⊙	⊙	☐
Help in obtaining financing	⊙	⊙	⊙	⊙	⊙	☐
Construction knowledge	⊙	⊙	⊙	⊙	⊙	☐
Keeping you informed	⊙	⊙	⊙	⊙	⊙	☐
Help in making decision to buy	⊙	⊙	⊙	⊙	⊙	☐

Please mark the ONE MOST IMPORTANT item affecting your satisfaction ──

Was the information you received reasonably accurate and were your questions before the decision to purchase satisfactorily answered? If no please comment: ☐ Yes   ☐ No

How would you rate your SUPERINTENDENT on each of the following:	Your Superintendent was					
	Excellent	Very Good	Good	Fair	Poor	
Informative at lot inspection	⊙	⊙	⊙	⊙	⊙	☐
Construction knowledge	⊙	⊙	⊙	⊙	⊙	☐
Informative at walk-through	⊙	⊙	⊙	⊙	⊙	☐
Maintaining quality	⊙	⊙	⊙	⊙	⊙	☐
Good at keeping you informed	⊙	⊙	⊙	⊙	⊙	☐

Please mark the ONE MOST IMPORTANT item affecting your satisfaction ──

(Continued)

FIGURE 3.2 *(Continued)*

Prestige Home Builders   Survey # 81

How would you rate our TRADE CONTRACTORS (subs) on each of the following:	Excellent	Very Good	Good	Fair	Poor	
Professionalism	◉	◉	◉	◉	◉	☐
Quality of work	◉	◉	◉	◉	◉	☐

Please mark the ONE MOST IMPORTANT item affecting your satisfaction ──

Please make any specific comments regarding trade contractors here:

_____

_____

Was the information you received at the builders meeting (where you met your superintendent and others) informative, accurate and understandable? If no please comment: ☐ Yes   ☐ No

_____

Did you experience any unexpected problems? If so please identify: ☐ Yes   ☐ No

_____

Are you pleased with your purchase thus far? If no please explain: ☐ Yes   ☐ No

_____

Would you confidently recommend our company to a friend or relative? If no please explain: If yes, please write their name and contact information here   ☐ Yes   ☐ No

_____

Please make any other observations or comments you feel would be helpful to our survey

_____

I prefer this information to be strictly confidential _____
You have my permission to use this for public display_____

_____   _____
Signature (optional)   Signature (optional)

- when drywall is up and taped (prior to painting)
- upon project completion
- others as required

Each inspection should be considered mandatory, and each is required by most building inspection departments.

Inspection Checklists

Many superintendents conduct internal inspections using checklists tailored to their particular organization. The checklists serve as memory joggers, prompting the superintendent to look for crucial items. Inspection checklists should include information about any items that:

- have been problems in the past
- are difficult to repair or replace if they are covered up
- are particularly important to quality in the eyes of the homeowners
- will cause a building code inspection to fail

Quality control checklists are invaluable tools for monitoring the quality and completeness of work. They can help ensure that minimum standards of quality are met. The quality control checklists in the Appendix to this book cover key stages in the construction process. Ideally, checklists should be developed and adapted for each activity, phase of construction or trade contractor. To make the best use of the checklists provided in the Appendix, adapt them to include measurable and attainable company-specific standards.

Preinspections

Inspections required by local building officials or departments, FHA/VA, and lending institutions are usually intended only to ensure compliance with standards of safety and structural integrity, not to deal specifically with quality control. Nonetheless, you should conduct preinspection checks to detect and correct quality problems before the inspector arrives. This worthwhile practice serves several purposes:

- it prevents rejection by the inspector
- it helps you avoid the embarrassment of failing an inspection
- it eliminates the cost of delays and reinspections
- it maintains your company's reputation as a high quality builder

Gaining Cooperation

Develop and promote a cooperative attitude during inspections. Extend courtesy toward inspectors whenever possible. Like most people, inspectors will themselves tend to be more cooperative under such circumstances. To develop an atmosphere of cooperation, follow these guidelines:

- Know the applicable building codes inside and out.
- Know what additional requirements have been established through the legal process by the local building department.
- Know what items or techniques inspectors would like to see but that are not official policies or standards.

- Schedule inspections following the procedures of the inspection department and as far in advance as possible, but don't order the inspection until the work is ready.
- Have trade contractors check their work to make sure it is ready.
- Check the job yourself to make sure it is ready before calling for an inspection.
- If a job is not ready when an inspection is scheduled, notify the inspector as soon as you are aware of a delay.
- Be available. Walk around with the inspector during the inspection whenever possible.
- Keep in mind that the inspector's interpretation of the codes can be as important as the letter of the code.
- Cooperate with the inspectors by promptly correcting work that is below standard.
- If the inspector requires something above and beyond the code requirement, find out why, and if the reasoning is sound, comply with the request.

Making yourself available is the most important key to passing your inspections the first time. If you are on the job when the inspector comes, you can correct most of the items immediately while the inspector is there. Once you have gained their confidence, more often than not, inspectors will take your word that you will correct many of the other items and allow you continue without a reinspection. Don't be defensive. If the inspector takes your word that an item will be corrected, make sure it is. The first time you fail to keep your word will be the last time you will have the opportunity.

If the inspector asks you to do something for which compliance seems unreasonable, try negotiating first. Refusing to comply may create hard feelings and can lead to considerable delays while both parties attempt to prove their points. Cooperation is always better than alienation.

If you experience a persistent problem with getting something to pass inspection and you feel you are correct in the way that you do something, work directly with the chief building official to get it resolved. If that doesn't work, get together with other builders in your local building industry association (BIA) or home building association (HBA) and flex some combined muscle to get the policy or requirement changed.

Making Corrections

All construction team members have the ultimate responsibility of ensuring that their work is equal to or better than the quality specified. However, the superintendent is in the best position to enforce these standards. Performance should be judged on the basis of compliance with the standards, and the individual or firm who completed the work should be responsible for any necessary corrective action.

Logging Inspections

All required inspections should be noted on the construction schedule and recorded in the daily logbook. The daily log should contain records of both in-house and outside inspections, including the following information:

- date and time of the inspection
- name of the inspector
- nature of the inspection

- results of the inspection
- notes on any special circumstances
- inspector's signature where applicable

Keep track of the inspection card! It is your only legal proof that an inspection has occurred and that the work passed the inspection. During inspections the card usually is required to be posted on the jobsite. Between inspections, however, you may want to consider keeping the inspection card in a more secure place.

Final Inspection

The final inspection before the homeowners' walk-through and orientation is one of the most important. Conduct your own personal inspection first. Make sure you are satisfied that the job is complete. You may want to have your immediate supervisor or another superintendent conduct this inspection. Ask this person to be very critical of even small discrepancies. Your familiarity with the job day after day may cause you to miss small flaws. Asking for another "set of eyes" and asking that person to be supercritical at this point will help you cut down on homeowner complaints and enhance your builder's reputation for quality.

Homeowners' Walk-Through and Orientation

If the proper inspections have already taken place, the homeowners' walk-through and orientation can be a positive, rewarding experience. Some builders even invite trade contractors to be present to demonstrate the operation and maintenance of their products. Trade contractors who have installed complex systems such as lawn sprinklers, HVAC, or security systems, are particularly well suited to educate the homeowners about how to use the systems. The trade contractors also can explain any warranties involved for their particular work and discuss how they will handle customer service requests. Their presence helps to make homeowners aware of their own responsibilities for maintenance and establishes a means of satisfaction for many homeowner complaints.

Instead of simply having the homeowners look for flaws or imperfections, accentuate the positive aspects of the home. (More information on the homeowners' walk-through and orientation appears in chapter 7.)

4

Cost Control

Completing a project within the projected budget is a primary goal of every superintendent. Every project should therefore have an established budget before a single nail is driven (or even ordered). This budget normally derives directly from the detailed estimate of the house or from accounting records of previously built homes of the same model. The budget typically breaks into two parts:

- a summary of the major costs involved in building the home
- a detailed account of all items, including the quantities and itemized costs for all the materials and labor required to build the house

The itemized costs allow you to track each item closely and take immediate corrective action whenever actual costs differ from the budgeted costs. The corrective action may be major or minor—some procedural or personnel change, or simply a recognition that the budget is in error and needs to be adjusted on future jobs.

Establishing the Budget

The way the budget will be subdivided will depend in part on the sophistication of the company and on the kinds of homes being built. For example, different work areas may be grouped as follows:

- job overhead
- sitework
- foundations
- masonry
- wood framing (labor and material, may be separate items)
- roofing and flashing
- exterior siding and brick
- heating and air conditioning
- plumbing
- electrical

- insulation
- drywall
- interior wood trim
- painting
- cabinets
- floor covering
- appliances
- landscaping, walks, and driveways

All builders should develop a detailed chart of accounts that reflects their own particular needs and type of construction. The chart of accounts should be consistent from one job to the next yet flexible enough to handle the different types of work the company performs, from new construction to remodeling projects. Lee Evans originally proposed an excellent sample chart of accounts for builders and remodelers in the 1960s. Over the years it has been modified and updated several times by the Business Management Committee of NAHB. The complete NAHB Chart of Accounts is found in NAHB's *Accounting and Financial Management for Builders and Remodelers* (see Additional Resources). All residential builders and remodelers should take advantage of this excellent source of information. It was developed by builders for builders and provides an excellent format for establishing a chart of accounts.

A well-tailored chart of accounts can be developed from the detailed estimate of the house, including construction materials, subcontract costs, direct labor, maintenance and repair expense, project overhead and general administrative or office expenses. Using the estimates for each category of work, set up budget line items in the same sequence as the steps of construction.

When subdividing the work areas, keep in mind that expenditures for each work item must be easily measurable. Keep items that occur within different time frames separate even if they are purchased from the same source. For example, it would be unwise to group rough grading and finish grading or framing lumber and trim material, even though these materials and labor may come from the same trade contractor.

Comparisons against the budget need to be made as quickly as possible after each delivery or activity is completed to see if the job is still within the targeted budget. Timely checking of expenditures against the budget will often allow you to take corrective steps before it is too late (see Figure 4.1).

It is important that the estimating and cost accounting records be as accurate as possible. Building companies are best served by using an integrated numbering system for accounting, estimating, purchasing, and other functions. If you have in-house employees who perform direct labor (carpenters, masons, and so forth), maintaining accurate timesheets relative to each work category and keeping up with materials moved from one job to another will be essential to determine whether or not a job is profitable.

Material Control

In order to control the cost of a project you must control the materials used in the construction of the home. Proper use of materials is crucial to any construction company's cost-effective operation and it is therefore an important superintendent responsibility. Material control is a function of the following basic building activities:

- value engineering and planning
- completing specifications

FIGURE 4.1 Sample Project Budget

Construction Company Customer Name: Phone Number:	Johnson, Dave (555)555-4545	Lot # & Subdivision: Bid Total:			38 Cedar Hollow		
Item Description	Estimated Cost or Bid	Draw 1	Draw 2	Draw 8	Total Draws	Variance	%Var.
Project Overhead							
Lot	$ 59,290	$ 59,290			$ 59,290	$ -	0.0%
Building permit and fees	$ 7,500	$ 7,623			$ 7,623	$ (123)	1.6%
Plans and engineering	$ 2,125	$ 2,125			$ 2,125	$ -	0.0%
Temporary utilities	$ 500	$ 321			$ 321	$ 179	-35.8%
Construction loan interest	$ 8,765	$ 8,651			$ 8,651	$ (114)	-1.3%
Supervision	$ 17,462	$ 16,789			$ 16,789	$ (673)	-3.9%
Contingency	$ 1,792	$ 275			$ 275	$ (1,517)	-84.7%
Subtotal of Project OH	$ 97,434				$ 95,074	$ (2,360)	-2.4%

Item Description	Estimated Cost or Bid	Draw 1	Draw 2	Draw 10	Total Draws	Variance	
Hard costs							
Earthwork	$ 4,350	$ 4,302			$ 4,302	$ (48)	-1.1%
Footings	$ 2,400	$ 2,467			$ 2,467	$ 67	2.8%
Foundation	$ 9,008	$ 9,008			$ 9,008	$ -	0.0%
Flatwork	$ 9,003	$ 9,889			$ 9,889	$ 886	9.8%
Miscellaneous steel	$ 1,756	$ 1,756			$ 1,756	$ -	0.0%
Window wells	$ 774	$ 774			$ 774	$ -	0.0%
Dampproofing	$ 187	$ 187			$ 187	$ -	0.0%
Utility laterals	$ 1,200	$ 1,450			$ 1,450	$ 250	20.8%
Septic tank	$ -	$ -			$ -	$ -	
Potable water well	$ -	$ -			$ -	$ -	
Framing material	$ 37,892	$ 39,987			$ 39,987	$ 2,095	5.5%
Framing labor	$ 11,175	$ 11,175			$ 11,175	$ -	0.0%
Entry doors	$ 1,750	$ 1,750			$ 1,750	$ -	0.0%
Garage doors	$ 1,358	$ 1,358			$ 1,358	$ -	0.0%
Windows	$ 4,623	$ 4,623			$ 4,623	$ -	0.0%
Plumbing	$ 11,256	$ 5,000	$ 6,546		$ 11,546	$ 290	2.6%
Heating	$ 10,693	$ 4,500	$ 6,193		$ 10,693	$ -	0.0%
Air conditioning	Included in HVAC				$ -		
Electrical	$ 6,462	$ 2,000	$ 4,645		$ 6,645	$ 183	2.8%
Lighting fixtures	$ 3,000	$ 3,578	$ (578)		$ 3,000	$ -	0.0%
Roofing	$ 3,488	$ 3,488			$ 3,488	$ -	0.0%
Insulation	$ 4,086	$ 4,086			$ 4,086	$ -	0.0%
Drywall	$ 13,165	$ 5,000	$ 8,165		$ 13,165	$ -	0.0%
Finish carpentry material	$ 4,293	$ 4,565			$ 4,565	$ 272	6.3%
Finish carpentry labor	$ 6,380	$ 6,380			$ 6,380	$ -	0.0%
Painting	$ 6,730	$ 6,730			$ 6,730	$ -	0.0%
Tile/marble	$ 8,335	$ 8,335			$ 8,335	$ -	0.0%
Fireplaces	$ 3,678	$ 3,678			$ 3,678	$ -	0.0%
Floor coverings	$ 8,690	$ 8,690			$ 8,690	$ -	0.0%
Cabinets	$ 21,000	$ 22,111	$ (1,111)		$ 21,000	$ -	0.0%
Countertops	Included in cabinets				$ -		
Hardware and mirrors	$ 2,556	$ 2,340			$ 2,340	$ (216)	-8.5%
Siding	$ -	$ -			$ -	$ -	
Soffit and fascia	$ 2,954	$ 2,954			$ 2,954	$ -	0.0%
Gutter	Included in soffit & fascia				$ -		
Exterior railings	$ 5,640	$ 5,640			$ 5,640	$ -	0.0%
Decks	$ 7,898	$ 7,898			$ 7,898	$ -	0.0%
Foundation plaster	$ 287	$ 287			$ 287	$ -	0.0%
Cleanup	$ 1,500	$ 1,456			$ 1,456	$ (44)	-2.9%
Subtotal of Hard Cost	$ 217,567	$ 197,442	$ 23,860		$ 221,302	$ 3,735	1.7%
Total Hard Cost & Proj. OH	$ 315,001	$ 197,442	$ 23,860		$ 316,376	$ 1,375	0.4%
Company Overhead (G&A)	$ 10,751	$ 10,751			$ 10,751	$ 10,751	0.0%
Profit	$ 18,987	$ 17,612			$ 17,612	$ 17,612	-7.2%
Total Sales Price	$ 344,739						

- ensuring accurate contracts
- purchasing
- scheduling deliveries
- providing proper storage and care of materials
- avoiding material waste and misuse

Introduction to Value Engineering

In the home building industry *value engineering*—sometimes called optimum value engineering (OVE)—is a systematic approach to evaluating the least costly method of building without sacrificing quality or function. Much of the industry's research in OVE techniques was done by the NAHB Research Foundation in conjunction with the U.S. Department of Housing and Urban Development (HUD).

OVE systems offer many useful and money-saving guidelines for the use and conservation of materials. In addition, when OVE principles are applied systematically and appropriately, they produce a better and more economical product—one that serves its function in many cases better than a house built in a more traditional manner. For example, studies have shown that placing studs and floor joists 24 inches on center and aligning them with the roof trusses (which are normally 24 inches on center) not only conserves materials but also increases the structural stability of the building (Figure 4.2).

The systematic approach of OVE involves a group of closely related cost-saving methods. It begins during the planning process, takes into account alternative construction techniques, and is carefully integrated with the construction process so that all phases of construction effectively complement each other. While OVE techniques do not require detailed engineering analysis, professional engineering advice may sometimes prove helpful in making the most of OVE concepts. Having an engineer design or evaluate each house, however, is unnecessary; concepts that prove valid on one house can be applied in similar situations to other jobs.

Value Engineering: Three Steps

Value engineering is based upon three simple steps:

- gathering information
- identifying, analyzing, and evaluating alternatives
- selecting and implementing the best alternatives

Gathering Information. In OVE it is important to first determine the dollar value of each item used in construction. In some cases an item may have aesthetic value but little or no structural value. The worth must then be determined by the individual purchasing the product. For example, a home buyer may wish to have a certain type of window placed in his or her home out of personal preference, perhaps because they find this window particularly attractive. This type of consideration is entirely different from purchasing windows to achieve energy efficiency.

Identifying, Analyzing, and Evaluating Alternatives. Identifying cost-saving techniques involves generating new cost-saving ideas to serve necessary functions within the house being built. For example, many of the newest, most innovative approaches focus on efficient floor plans and products that make the most of fewer square feet while producing a home that meets the needs of potential buyers. Analyses of new approaches should be both systematic and objective. In order to determine the

FIGURE 4.2 Comparison of 16-inch and 24-inch On-Center Framing

**Standard framing
16 inches on center**
Truss, stud, and joist may
or may not be in line.

**OVE framing
24 inches on center**
Truss, stud, and joist are
directly in line. Load is
supplied more efficiently.
In addition to savings on
top and bottom plates
and band joist, structural
headers may be omitted.

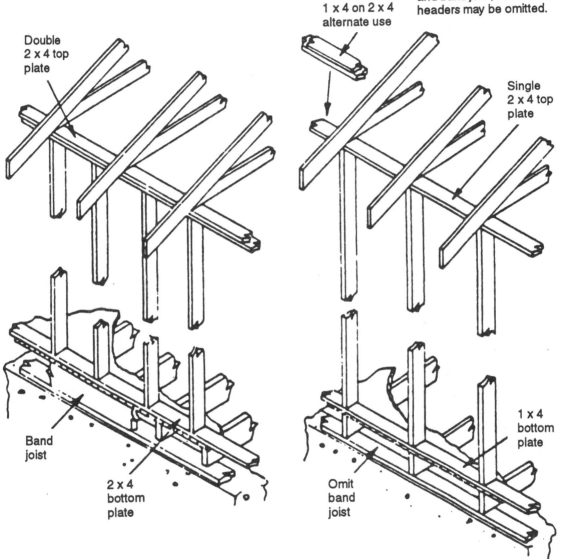

Double
2 x 4 top
plate

1 x 4 on 2 x 4
alternate use

Single
2 x 4 top
plate

Band
joist

2 x 4
bottom
plate

Omit
band
joist

1 x 4
bottom
plate

advantages and disadvantages of an alternative approach, each approach must be evaluated and tested to ensure feasibility.

Keep in mind that homeowner acceptance must be a key factor in determining feasibility. It does no good to develop a newer, cheaper, better way of doing something if it is not what the consumer wants or is willing to buy. As the superintendent, you must evaluate alternatives from the homeowner's point of view and not your own. There are many construction methods and materials that the builder may prefer to use but that may be unacceptable to the homeowner. A simple way to estimate the value of an alternative is by asking: What does it cost overall and what are the potential savings? Other factors to consider when evaluating alternatives include aesthetics, durability, marketability, and lifetime maintenance cost and convenience.

Selecting and Implementing Alternatives. As alternative approaches are evaluated, some ideas will clearly fail to meet a home's functional requirements and will be rejected outright. Other alternatives may appear to have great potential but require additional information before reaching a final decision. Reserve such alternatives for further research. Select those ideas that offer the greatest savings while still maintaining the functional qualities desired. Once a final decision has been reached, record and monitor all costs, including installation, maintenance, and use of the product or technique. Monitoring is important to ensure that the anticipated savings are realized.

Many builders who have implemented OVE and taken steps to reduce costs without sacrificing quality indicate that they have saved between $1,000 and $20,000 per home.

Complete Specifications

Many homes built in the United States are constructed without detailed written specifications. At some point, however, you must decide the quantity and quality of all materials and which trade contractor will perform the work. You may be provided with minimal plans and specifications and be asked to construct a house from sketchy information. It is important to establish specifications up front, make these decisions, and set the expectation levels of the homeowners before you begin construction.

Detailed plans and specifications can eliminate many problems before they arise. By clarifying the homeowners' desires (in the case of custom work) and completing the specifications, builders and superintendents can improve jobsite efficiency. They also can make intelligent, accurate decisions when submitting bids for custom work.

Regardless of who furnishes the plans and specifications, make certain that every item is specified or that at least an allowance is agreed upon. When setting allowances it may be tempting to recommend low values in order to make the overall price more attractive. However, setting realistic initial allowances helps keep the homeowners happy by minimizing the possibility of later surprises and their accompanying higher costs. This professional approach will help to keep your company's good reputation intact.

Ensuring Accurate Contracts

Legally, anything that is not part of the contract is not required of the builder. In practice, to maintain good working relationships with their homeowners many builders will give in to the homeowner's wishes rather than face an argument. Keep in mind, however, that if a buyer has contracted to buy a production house, you should not

Purchase Orders

One custom home builder recently related an interesting experience that he had with the use of purchase orders. Before this builder started using written purchase orders, at least one-third of his superintendent's time was spent running around town picking up materials. The builder said, "He was spending between 150 and 200 hours chasing items per house. It seemed he was always away from the job going to get something." Since implementing the purchase order system the builder has found the superintendent "on the job all the time. He rarely has to go chasing after anything."

This builder also noticed that written purchase orders reduced the number of homeowner changes. "Once I explained to my customers that all of the material would be purchased up front with the use of purchase orders, they somehow realized the importance of making decisions early and sticking to them." The builder now takes whatever time is needed for customers to make firm initial selections. Decisions are made up front regarding what will be needed and when, so materials are on the job when we need them. Then, "once we start construction everything (and I mean everything) flows more smoothly. I'll bet the number of change orders has been cut by 75 to 85 percent." The builder's customers like this system better, too. They don't have to take time off work to go look at items they want to change or pick out tile colors and light fixtures.

be forced to deliver a custom house at the same price. Specifying what is wanted up front and in writing helps home buyers and builders reach a firm agreement. Contracts should be detailed enough to describe what is expected of each party. Fortunately, you need not fill the contract with a lot of legal or technical language. A good attorney should review your contracts to make sure both parties are adequately protected but simple is usually much better.

The same principles hold true for your contracts with trade contractors. The builder should initiate a written subcontract agreement with each trade contractor. Again, complex legal language is unnecessary. Simple agreements written in your own words are fine as long as they are clear (and reviewed, if not written, by your attorney). Each agreement should outline what you expect and what you are willing to accept from your trade contractors. It should include details such as how and when you expect to make payment, cleanup procedures, and OSHA and building code compliance. The NAHB has publications on contracting that include sample contract language, as well as sample subcontract agreements (see Additional Resources).

Purchasing

Building houses without purchase orders is like framing houses without pneumatic nail guns. It can be done, but it is really hard to make money. An effective purchasing system based on written purchase orders can not only save builders a lot of money but also reduce the time your workers and trade contractors spend running around town picking up materials that either were not ordered or have not yet been delivered to the site.

Studies show that by implementing a good purchasing system with formal purchase orders for all material and subcontracts, builders can raise their profit margin by several percentage points. One particularly well-managed company implemented a purchase order system and increased its bottom-line profit margin from 9 percent to over 16 percent in one year. That's a 78 percent increase in profits in one year!

Superintendents and the trade contractors under them are in the best position to monitor day-to-day material needs, including when specific materials are needed and where they should be placed on the jobsite. While office staff may initiate purchase orders and send them to trade contractors and suppliers to authorize purchases, the superintendent should place the final will-call order and request delivery (Figure 4.3).

The first step in applying a purchase order system to a job is to complete a detailed estimate created for the home. The estimate must contain absolutely everything required for building. From a completed estimate, purchase orders can be generated for everything from the excavator to the floor coverings. Purchase orders can be automatically generated and printed from many computerized estimating systems.

The next step is to review all of the purchase orders. If an estimator prepares the estimate, he or she may not know exactly what happens on the jobsite or specifically how the home will be built. Because the superintendent is responsible for building the

FIGURE 4.3 Purchase Order System Paper Flow

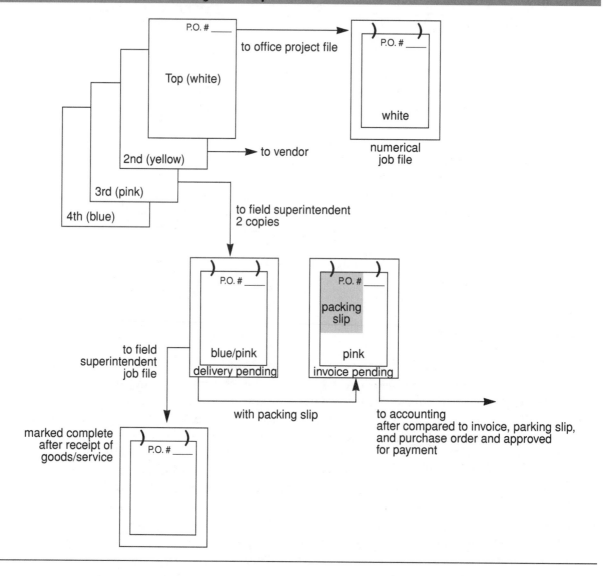

job and coordinating the trade contractors, you need to review the estimate to make sure that the purchase orders cover everything needed to build the home and that the proper trade contractors have been assigned to the job. The approved purchase orders are then mailed to all trade contractors and suppliers before construction begins. Some builders enclose other helpful documents with the purchase orders. One medium-sized builder who works on scattered sites throughout his state includes for each trade contractor the necessary copies of the plans and a map to each job.

Sending out purchase orders in advance has many benefits, including the following:

- Once accepted, the purchase order becomes a legal contract that obligates the vendor to supply the material or labor specified at the price indicated.
- The vendor is notified that you need the material specified and will generally reserve materials in short supply for your particular job.
- Prices are pretty firmed up and in many cases guaranteed. Even when there are price increases during construction, most suppliers and trade contractors will hold the prices specified in the purchase order. Because the purchase order is a contract, the vendor will adhere to the price more readily than if you had only requested an estimate. In essence you have already agreed to purchase the material. You simply are delaying delivery until the proper time.
- Your trade contractors all know what you are going to build far in advance and they can begin to plan schedules and teams accordingly.
- The purchase of materials and supplies is limited to only those authorized by the original estimate or purchase order, eliminating waste.
- Items not included in the original budget can be easily tracked (see Handling Budget Variances).
- Ordering and delivery of materials is much simpler. You just request delivery of a particular purchase order.

When the material is delivered, the superintendent or trade contractor checks the delivery to make sure the material is complete and is of the appropriate quality. When the trade contractor completes the work it is approved for payment. In many cases it is not even necessary to have the trade contractor or supplier submit a separate invoice or statement. The superintendent approves the material or subcontract work for payment and the accounting department makes payment directly from the purchase order. The purchase orders eliminate the need for invoices and unnecessary paperwork and the trade contractor or supplier often is paid more quickly than would have been the case using an invoice system. Many national builders now use this kind of payment system and after a little training their trade contractors and suppliers have found it to be very efficient and beneficial.

Handling Budget Variances

A variance is any cost that is not included in the original estimate or budget. Obviously variances should be avoided. You can reduce budget variances if you:

- build well-designed homes with complete plans and specifications
- obtain accurate estimates of material and labor
- maintain a good purchase order system
- schedule just-in-time material deliveries
- count and inspect materials for quality and quantity

- train trade contractors to use materials properly
- use cut sheets to show trade contractors how you propose to use materials
- negotiate long-term pricing for materials and labor
- maintain proper security on the job to prevent theft and vandalism
- protect materials from extreme weather
- reduce or eliminate change orders
- closely supervise quality on the job
- protect previously completed work

We all know that it is almost impossible to build a house without any variances. When variances do occur it is important to track them and identify the cause of each one in order to reduce their occurrence in the future. The purchasing system accommodates this need with *variance purchase orders* (VPOs). Where a purchase order authorizes the purchase of materials and subcontracted work included in the original estimate, a VPO authorizes purchase of materials or work *not* included in the original estimate. The concept of the variance purchase order system has been promoted for many years by Lee Evans. While a VPO system requires a fair amount of discipline, the resulting savings on the purchase of unwarranted items or subcontracted work make it worth the effort.

Fill out a VPO for any material or work that was not included in the original estimate. In most cases, the VPO is filled out by the superintendent in the field and authorizes immediate delivery of materials or performance of work. A VPO typically contains four sections:

- identification of pertinent job information
- listing of the materials or work to be performed
- explanation of the cause of the variance, including a variance code
- identification of the action needed to avoid the variance in the future

The completed VPO is then approved for payment by the superintendent's supervisor. The variance codes on a VPO allow you to track the causes and amounts of variances on each job and to compare them from one job to another, from one trade contractor to another, and even from one superintendent to another. If you

---

VPOs Lead to Savings

One builder noticed excessive costs for tow trucks on his variance summary report. An abnormal amount was being spent pulling stuck delivery trucks out of the mud. After gathering more information the builder noticed that the number of stuck trucks was excessive on homes with a side-load garage. The builder always installed a temporary gravel driveway to allow delivery trucks access to the site. The trucks were getting stuck when they got off the temporary driveway to deliver materials close to the front door.

The builder decided to install a short spur off the temporary driveway over to the front door during the rainy season. The gravel drive went across the front yard to the front door. The short spur was removed when the final grading was done in preparation for the permanent driveway. The spur was not costly and virtually eliminated tow truck costs. The builder never would have known the extent of the problem—or found a solution—had he not been paying attention to the variance summary report.

can identify the common causes of variances, you can take the necessary steps to avoid them—or at least reduce their number and dollar amount.

A computerized variance summary report also allows you to identify trends. A variance summary report presents the aggregate cost of each type of variance over a particular period of time (month, quarter, year—see Figure 4.4). Using the cost codes on the VPO, you can determine the total of all variances resulting from bad weather, from superintendent, trade contractor or estimating errors, from engineering problems, and so forth. When you know the relative causes of variances you can focus your efforts most efficiently to prevent future cost overruns.

Variance Analysis

Superintendents need feedback on how they are doing with regard to the budget. An effective way to communicate the status of the budget is through the use of a variance report (see Figure 4.5). A variance report presents the original estimate's cost and the actual cost for each cost category, and highlights the differences. The superintendent or any of the other managers can tell at a glance how they are doing.

Some builders provide a copy of the weekly superintendent variance report to each superintendent as part of their weekly production meeting. Because superintendents know on a weekly basis where they stand on each home, including which projects might need adjustments, they find it easier to concentrate their efforts on bringing errant projects back into line.

Scheduling Deliveries

Most of what you need to know about scheduling deliveries is just common sense: materials should be delivered when they are needed and stored close to where they will be used. Proper scheduling of material deliveries cuts down on wasted time, money, and materials. Just-in-time delivery is very efficient and can save money by reducing waste. Materials do not sit around and get damaged, warped, weathered, or dirty. They can be placed in a convenient location and do not have to be moved out of the way. They can be bundled conveniently for efficient use, and quality of materials can be more effectively monitored.

Try to avoid scheduling material deliveries just before weekends; most jobsite theft occurs on weekends and holidays.

Have a Specified Delivery Location. Identify each project with a unique job number and post a job sign containing the number in a conspicuous location on the site. Putting the same job number on all delivery tickets for a particular job helps to ensure that materials arrive at the correct location.

Important: Have delivered materials stacked with the materials to be needed first on top of the pile. One way to accomplish this arrangement is to list those items required on top of the pile at the *bottom* of the purchase order. Material suppliers usually load their trucks starting at the top of the list. Materials on the bottom of the purchase order will likely be loaded onto the truck last and therefore end up on the top of the pile. When you send the purchase order to the supplier you may want to include a simple plot plan showing the exact location on the lot where the materials should be placed. Identify the location on the site with a small, simple sign, such as "Lumber Drop Here." Most drivers will deliver the materials to the location indicated and workers will not waste time trying to find what they need.

FIGURE 4.4 Annual Variance Report

Variance Code	Variance Code Name		$ of Variance	% of $ Total	# of Variances	% of # Total
11	Standard print error		$875.05	0.94%	7	0.40%
12	Special plan error		1,053.14	1.13%	20	1.14%
		Total Plan Variances	1,928.19	2.07%	27	1.54%
14	Code-required change		1,723.33	1.85%	34	1.94%
		Total Code Variances	1,723.33	1.85%	34	1.94%
23	Price		1,029.44	1.10%	36	2.05%
24	Incorrect material		2,095.18	2.25%	30	1.17%
21	Omission		2,013.36	2.16%	68	3.88%
22	Quantity		2,786.66	2.99%	86	4.91%
		Total Estimating Variances	7,924.64	8.50%	220	12.55%
45	Freezing damage		1,489.71	1.60%	23	1.31%
96	Weather precaution		1,166.90	1.25%	75	4.28%
41	Rain/wet site		6,570.97	7.05%	59	3.37%
43	Wind/storm damage		3,036.36	3.26%	31	1.77%
		Total Weather Variances	12,263.94	13.15%	188	10.72%
33	Bad lot access		3,122.38	3.35%	15	0.86%
32	Poor soil discovered		2,418.02	2.59%	8	0.46%
		Total Site Variances	5,540.40	5.94%	23	1.31%
51	Theft		2,715.21	2.91%	41	2.34%
		Total Theft Variances	2,715.21	2.91%	41	2.34%
62	Price change		6,753.30	7.24%	181	10.33%
61	Inferior material replaced		387.13	0.42%	54	3.08%
66	Delivered quantity		1,250.33	1.34%	90	5.13%
65	Late delivery		455.55	0.49%	8	0.46%
		Total Material Variances	8,846.31	9.48%	333	19.00%
76	Misuse of material		44,797.70	5.14%	73	4.16%
72	Trade contractor damage		2,286.76	2.45%	60	3.42%
71	Inferior work		8,868.96	9.51%	143	8.16%
75	Late trade contractor		2,055.98	2.20%	28	1.60%
74	Price change		1,179.81	1.26%	16	0.91%
		Total Subs Variances	19,189.21	20.57%	320	18.25%
81	Design change		2,088.87	2.24%	24	1.37%
87	Management error		1,965.22	2.11%	28	1.60%
82	Engineering change		69.15	0.07%	2	0.11%
91	Construction manager error		5,872.56	6.30%	213	12.15%
92	Field error		11,754.26	12.60%	213	12.15%
		Total Management Variances	21,750.06	23.32%	480	27.38%
86	Change to satisfy customer		11,387.48	12.21%	87	4.96%
		Total Customer Variances	11,387.48	12.21%	87	4.96%
		Grand Total	$93, 268.77		1,753	
97	Change orders		$17, 952.16	19.25%	147	8.39%
98	Keying error		$483.25	0.52%	1	0.06%
			$18,435.41	19.77%	148	8.44%

FIGURE 4.5 Variance Summary Report

Superintendent: John Christensen
Date of Report: 9/12/XX

Job Number	Customer	Start Date	Complete Date	Direct Cost	Variance Budget	Actual Variance	Balance	% Variance	Construction Time
974	Moore	3/3	6/18	$86,788	$867.88	$5,225.65	$4,358.77	6.0%	107
977	Kornack	3/13	6/24	92,654	926.54	7,442.49	6,515.95	8.0%	103
992	Leas	4/6	6/24	125,366	1,253.66	1,684.97	431.31	1.3%	79
993	Holt	4/15	7/4	158,023	1,580.23	1,988.74	408.51	1.3%	80
994	Nyhart	4/22	7/15	88,254	882.54	199.87	(682.67)	0.2%	84
995	Hayden	4/29	7/18	65,278	652.78	1,225.87	573.09	1.9%	80
996	Priest	5/4	7/24	111,253	1,112.53	357.00	(755.53)	0.3%	81
999	Kaiser	5/8	7/26	96,552	965.52	1,190.26	224.74	1.2%	79
1002	Cost	5/20	8/2	88,654	886.54	484.32	(402.22)	0.5%	74
1003	Mitchell	5/21	8/6	125,454	1,254.54	934.28	(320.26)	0.7%	77
1006	Stokes	5/26	8/15	102,525	1,025.25	219.74	(805.51)	0.2%	81
1013	Rayburn	5/31	8/20	108,789	1,087.89	411.70	(676.19)	0.4%	81
1016	Reed	5/23	8/25	75,988	759.88	811.67	51.79	1.1%	94
1019	Ulbrich	6/15	8/30	99,326	993.26	29.59	(963.67)	0.0%	76
1026	Prince	6/20	8/31	90,236	902.36	79.01	(823.35)	0.1%	72
1028	Galiher	6/13	9/2	82,659	826.59	644.32	(182.27)	0.8%	81
1029	Tuttle	6/22		92,888	928.88	355.25	(573.63)	0.4%	
1030	Johnson	7/13		79,698	796.98	788.00	(8.98)	1.0%	
1032	Fernandez	7/17		122,122	1,221.22	655.00	(566.22)	0.5%	
1033	Harmer	7/28		88,562	885.62	202.00	(683.62)	0.2%	
1034	Fielding	8/11		100,568	1,005.68	78.00	(927.68)	0.1%	
1038	Nish	8/21		86,555	865.55	0.00	(865.55)	0.0%	
Average				$99,862.44	$998.62	$1,433.09	$434.47	1.5%	83.06

Conducting Delivery Inspections. Always inspect materials on the jobsite immediately upon delivery. Problems often occur when deliveries are made to a jobsite where no one is present to direct the driver and immediately inspect the materials. Whenever possible, the superintendent or another responsible worker should be available at the site when a delivery is due. If for some reason this is not possible, direct the trade contractor who will use the material to check each load before it is disturbed in order to make sure it is complete and in good condition.

Manufacturers inspect many materials (lumber, for example) before shipment, but many latent defects can occur by the time materials reach the jobsite. For example, lumber can warp, crack, or split during shipment, making it difficult or impossible to use. In keeping with lumber grading rules common in the industry, as much as 5 percent of lumber in a delivery may be out of specification. While some of this lumber might be used as blocking, much of it cannot be used for anything and it should be returned. A thorough inspection by the superintendent can discover and prevent the use (and subsequent rejection) of defective materials. Rejecting an unacceptable floor joist before installation is much easier and less expensive than having to tear it out and replace it after installation. Clearly mark rejected materials and place them away from other materials to prevent confusion and accidental use.

Jobsite inspection of materials on arrival also uncovers any defects resulting from storage, transportation, or unloading. Proper inspection up front alleviates the problem of arguing with suppliers later over any damage. (If you are challenged by a supplier, keep in mind that suppliers often find themselves the victims of unjust claims for defective materials when, in fact, the problem occurred as a result of product misuse or abuse. Suppliers have consequently become very sensitive about complaints concerning damaged materials. A supplier is more likely to trust and accept a claim for defective materials if it is based on prompt inspection upon delivery, as following this practice limits the room for blame on both sides.)

Most suppliers are honest and will not intentionally short a builder. However, employees are human and sometimes forget a particular item or leave materials back at the warehouse instead of delivering them to the jobsite. This problem is common for small items such as nails, bolts, and adhesive, particularly in large orders where small items are easily overlooked. Inspect each shipment carefully to ensure that all materials have been delivered as ordered.

Finally, always be conscious of your right and your obligation to reject faulty materials. Most material suppliers will stand behind their products and substitute good materials for inferior ones, usually at no cost to you.

Providing Proper Storage and Care

Materials on the jobsite are easily damaged and wasted because of improper storage and care. For example, materials can be ruined by mud or snow during extreme weather conditions. In hot weather, common to many parts of the country, exposed lumber can dry out and warp. When shipped to hot, dry climates, even kiln-dried lumber can be ruined through careless storage.

All materials subject to deterioration should be covered at all times. A bundle of lumber should be used as soon as possible after it is opened to prevent warping and twisting. Trim and other finish material should not be stored on concrete floors. The finish material absorbs moisture from the fresh concrete and may warp. Twenty guidelines for the conservation and care of building materials, prepared by Lee Evans in his book, *Quality in Construction,* are offered in Figure 4.6.

FIGURE 4.6 Guidelines for Conservation and Care of Building Materials

1. Can materials be ordered in smaller quantities to reduce exposure time?

2. Can the supplier load them so that materials used first are on the top instead of the bottom of the load? (Because of the time saved, builders may be able to afford to pay something extra for this service, especially if they use their own crews or can negotiate with trade contractors to recover some of the time savings.)

3. Can the superintendent do anything else to improve planning load composition, size, or delivery spotting to avoid extra handling?

4. Can any kind of temporary cover such as plastic be used? (Provide shelter but do not make a sweat box by covering materials too tightly.)

5. Can the load be banded in several batches rather than in one big load?

6. Is a spot by the foundation prepared for the load? Does it provide drainage? Are skids provided to keep the load off the ground? (Ideally it should be up 6 inches, and in bad weather, polyethylene laid below the load to prevent moisture rise.)

7. Are asphalt roofing materials stored flat? (Curved and buckled shingles create an unsightly roof.)

8. Can any of the load delivery spots be covered with stone or paved? (In multifamily projects paving or stone is desirable because subsequent handling can be done with lift trucks, but doing that requires much planning and scheduling.)

9. Are siding materials stored so that they will not be scratched or damaged?

10. Can scheduling be changed to prevent deliveries from being exposed to weather for more than a day or two?

11. Can theft of site materials be reduced? Can materials be nailed down or locked up before dark? Will reducing deliveries to one-day requirements help? Can materials be banded, wired, nailed together, or marked to help recover or identify them for prosecution of a thief? Can superintendents and builders cooperate to do anything in the way of rewards for information about stolen materials?

12. What can be done to prevent vandalism?

13. Can superintendents train in-house labor and trade contractor workers to take better care of materials? (Stacks that are torn into and scattered deteriorate rapidly.)

14. Is too much material ordered for a particular unit or building resulting in waste, downgrading, or rehandling?

15. Is a place prepared for temporary storage of trim and similar materials so that they are dry, straight, and not disturbed?

16. Is each unit dried-in quickly enough? Is siding applied soon enough to prevent water drainage in the house?

17. Is temporary drainage provided and directed away from the house?

18. Can some deliveries and phases of construction be rescheduled so that materials are not damaged by workers? (For example, lay flooring materials late in the process. Don't deliver doors, doorframes, and windows until the unit is ready for installation. These items, as well as finishes, are damaged when workers move them.)

19. Can protective coverings be provided for such items as bathtubs, flooring materials, plastic laminate tops, and appliances?

20. Can temporary furnace heat be provided to help dry out the house and to prevent moisture buildup in flooring materials, trim, and paneling?

Source: Adapted from Lee S. Evans, *Quality in Construction* (Washington, DC: National Association of Home Builders, 1974) p. 51.

Avoiding Material Waste and Misuse

The large scrap piles found on many construction sites are the most obvious symptom of the growing problem of material waste in the construction industry. Taking the time to preplan your use of materials can eliminate a substantial amount of this waste. Spend a little extra time at the drafting table and graphically depict how materials are to be installed. The resulting drawings can communicate to your workers and trade contractors the exact manner in which materials are to be used. This eliminates waste and saves money. Many builders also distribute copies of approved construction methods and procedures further stipulating the efficient use of materials. The production manual discussed earlier is ideal for communicating expected standards of production.

Giving your framer a copy of the estimate (without prices) also is a good practice. The estimate normally indicates the type of material to be used and its intended use. For example, a 2x4x16' board could be used for myriad things ranging from wall plates to false fascia. How is a trade contractor or framer supposed to know what all of the materials in a pile of lumber are designed for without a copy of the estimate? By referring to the estimate, the carpenter will know that the 2x4x16' is to be used as a fly rafter—not backing material.

Excess material is not waste. Excess material should be protected, stored, and returned for credit as soon as possible. If you do a good job of estimating there should be very little excess material, but there are substantial savings available to the builder by returning excess materials for credit. Sometimes you will have to pay a slight restocking charge, but the overall savings are worth it.

Return Excess Material for Credit

One builder has his framing lumber separated into four separate deliveries. The last load contains the roofing felt and drip edge. This builder has a standard policy that all excess materials left over from the rest of framing are to be set aside and tagged so that the delivery truck driver can pick up the materials and return them for credit.

The material is specially tagged and covered with the wrapper from one of the lumber packages. Over time, the truck drivers have been trained to look for the excess material on the roof load; the builder doesn't even have to call and tell the lumber supplier that there is a pickup ready for return.

Labor Cost Control

Many builders have focused a major portion of their cost control efforts on labor costs as a primary avenue for potential savings. This greater direct control of labor may have been possible in the past when builders employed crews for large portions of the work. However, with the long-standing trend toward the increased use of trade contractors, labor efficiency is now largely under the trade contractor's control. Of course, the superintendent who discovers ways for a trade contractor to increase productivity may lower costs for future jobs. In addition, you can help to keep costs down by watching carefully for trade contractor errors that may be properly backcharged to the responsible trade contractor. Backcharging a trade contractor can be a tricky process, however. As often as not, a backcharge on one job becomes a price increase in the next job. It has been the experience of many builders that backcharges simply do not work very well. It is more effective to educate trade contractors and cooperate with them than to blame and alienate them.

5

Scheduling

Completing each job on schedule is another primary goal of the superintendent. With so many interrelated functions that must be accomplished in proper sequence, construction must progress in the most efficient manner possible to be economical. Effective scheduling drives this efficiency and contributes directly to the excellence of any successful builder in today's competitive marketplace. Many contracts are awarded on the basis of the builder's ability to complete the project in a timely manner.

Any builder will tell you it is easier to build a house with a good written schedule than it is to do so without one. A formal, written schedule allows you to build a house with less effort because it organizes the work sequence, much as a good set of plans and specifications organizes what is to be built. As a result, you can build more houses with less effort than when using informal scheduling methods or no scheduling method at all.

Why Schedule?

If you still wonder why you need to adopt a more formal scheduling approach, consider the following reasons:

- Scheduling is one of the two most important things you need to do before you start a project. Most successful builders would not begin a project without creating a detailed estimate of all of the things that contribute to the cost of a project. It is just as essential to plan out the sequence and interaction of all of the activities that must be performed to complete the home.
- Formal scheduling allows you to build the project in your head before you start incurring expenses for materials and labor. If you can build the project in your head you will also likely be able to anticipate, prevent, or solve most of the problems that might arise with building that home.
- A well-prepared schedule allows you to notify the various trade contractors well in advance of when they are needed so they can plan their own schedules. If

trade contractors know in advance when you will need them they are much more likely to be on time. Shared with the trade contractors, the schedule can foster the effective two-way communication that is essential to successful contractor management.

- Scheduling can help to level out the up-and-down cycles typical to the construction industry by reducing slack time and increasing overall productivity.
- A thoroughly prepared project schedule will highlight bottlenecks where labor, equipment, or materials are too tightly scheduled and also reveal activities or time periods for which resources are spread too thin.
- Formal scheduling provides greater flexibility in work assignments, yielding more options when unexpected delays occur as a result of equipment breakdowns, trade contractors failing to show up, or bad weather.

The biggest reason to use formal scheduling systems is because they work! Formal scheduling is a very powerful tool that allows you to manage projects more effectively.

Getting Started

It is possible for a superintendent to use a schedule effectively without being able to create or draw one. However, knowing how to create a formal schedule on paper should help you to use and understand the schedule more effectively. This chapter provides an overview of basic scheduling methods and their benefits and describes many of the steps necessary to create a project schedule. Superintendents and builders can find more information about formal scheduling procedures in the Home Builder Press publications *Scheduling Residential Construction for Builders and Remodelers* and *Bar Chart Scheduling for Home Builders* (see Additional Resources).

Scheduling Methods

The type of scheduling system a building company uses depends on several interrelated factors:

- size of the building company
- volume of work
- type of construction
- owner or architect's requirements (in custom work)
- project location
- competition
- project size
- project complexity
- extent of subcontracted work
- capacity of the superintendent
- workload (current and anticipated)
- past experience with schedules
- contract provisions
- computer capability

While builders today use many scheduling methods and tools, including a variety of computer applications, this discussion will concentrate on two of the most practical scheduling options for either manual or computer use: the bar chart and the Critical Path Method (CPM).

Scheduling Phases

Formal scheduling for both the bar chart and CPM methods involves the following stages:

- planning, during which the project is broken down into different phases and activities
- scheduling, during which a duration (an allowance of time) is established for each individual activity or task
- sequencing, during which the activities' various interrelationships are defined and their sequence determined
- communicating, during which the overall schedule is put in written or graph form and communicated to the various participants
- monitoring, during which the superintendent uses the written schedule to manage the project
- updating, during which the schedule is continually revised as the project progresses

Sequencing Activities

Work should flow without interruption. To achieve this flow, however, activities must be properly sequenced from the beginning. Keep in mind that two or more distinct parts of a particular task occurring at different times in the job need to be scheduled as separate activities. For example, the electrical work should be divided into electrical rough-in and electrical trim-out.

An activity is a single task or step that has a recognizable beginning and end and that requires time to be accomplished. The various activities involved in a single construction project can be organized using the following categories:

- area of responsibility or craft
- structural elements
- location on the project
- materials vendor

Activities can be organized according to the trade contractors who will perform them. Keep in mind that trade contractors may appear more then once in the schedule if they perform more than one activity or if they perform work at different stages. Trade contractors who perform more than one task also may appear more than once on the schedule if their activities represent different structural elements completed at different times.

Similar activities, such as interior concrete, floors, and exterior flatwork may be considered separate activities because they are performed in different locations, usually at different times. Deliveries of materials supplied by different vendors (such as garage doors and interior doors, or trusses and other framing lumber) usually are considered separate activities.

Your final list of activities should be arranged in approximate sequential order. To develop an accurate sequence, for each activity ask yourself three crucial questions:

1. Which activities, if any, must precede this activity?
2. Which activities, if any, must follow this activity?
3. Which activities, if any, can be conducted simultaneously?

Constraints also affect the sequencing of activities. Typical constraints include the following:

- physical or logistical factors, such as availability of labor or equipment, construction methods, and safety constraints
- practical budgetary or safety considerations
- managerial requirements

Scheduling constraints may result from the homeowner's desires or requirements, the company's financial requirements, computer or accounting requirements, the availability of competent managerial control, or simply managerial preference. Be prepared to work within the guidelines established.

Determining Activity Duration

Most superintendents rely on their experience and knowledge of local resources and conditions when they assign durations to construction activities. Smart superintendents begin by obtaining a time estimate from each trade contractor. Sometimes, busy trade contractors may commit a token workforce to begin work on your jobsite while finishing up work for another builder. At other times, extra workers may be added to try to rush an activity or catch up a schedule. A good estimate of time requirements up front (and insistence by the superintendent that a full crew arrive to do the work at the appointed time) should help to curtail these problem behaviors.

To estimate accurate durations for construction activities, follow these rules:

- Evaluate each activity independently of all others.
- Obtain estimates for durations from trade contractors.
- Assume a normal workforce for your crew and for trade contractors; avoid overloading your workforce in an attempt to shorten the necessary activity duration.
- Assume normal production rates for the time of year in which the work will be performed.
- Assume a normal workday. If needed, overtime or multiple-shift work can be entered into the process later. While a certain portion of the work may be accelerated for a limited time to meet a crisis or solve a particular problem, this solution should be the exception rather than the rule.
- Use consistent time units, preferably full working days. Don't schedule some activities in days and others in hours. For the purpose of construction scheduling, the minimum duration of an activity should be one day. Also, use working days or calendar days consistently; don't mix them.
- Be as accurate as is practicable. Do not overestimate activity durations in an attempt to make the schedule more manageable or underestimate them in order to keep the schedule tight. Work tends to take as much time as you allow, so avoid putting too much padding into a schedule.

The Bar Chart

Scheduling with a bar chart allows all those involved in the construction process to see clearly where their task falls on the construction time line.

Planning the Bar Chart

The bar chart is a very basic scheduling method, requiring only that all construction activities be sequenced and that a duration be estimated for each.

Scheduling with the Bar Chart

A bar chart organizes construction activities by the calendar. The sample schedule in Figure 5.1 shows a complete list of activities with projected time slots. The bars on the calendar portion of the chart indicate when the activities are to occur. According to the chart, this simple house should be completed in 52 working days.

One of the biggest advantages to a bar chart is its simple visual clarity. You can easily see and understand when an activity will begin and end. The superintendent, workers, and trade contractors can see at a glance how the work has been planned to progress.

Monitoring the Bar Chart

The strengths of the bar chart as a planning and communication tool can limit it as a monitoring or management tool. A bar chart cannot show the complex interdependence of various activities.

In construction some activities can be completed relatively independently of other activities, so the builder or superintendent has great flexibility in scheduling them. For example, landscaping can be completed as early as the brick or siding and exterior concrete are complete or wait until just before you complete the house. Other activities absolutely must happen at a particular time or in a particular order lest the project be delayed. These activities are called *critical* activities. Framing, for example, is almost always a critical activity. A bar chart will not normally show the float, or slack time, in non-critical activities. By the same token, a bar chart does not show what happens when a particular activity is delayed. Suppose, for example, that delivery of the burgundy tub you ordered for a custom home has been delayed. What other activities are affected? A bar chart will not reveal how a delay in one activity will affect other activities. Therefore, you may wish to reserve the bar chart method for use as a general planning and communications tool when a graphic display of the interrelationships is unnecessary.

The Critical Path Method

Unlike the bar chart scheduling method, the Critical Path Method (CPM) identifies those activities that must be completed on schedule in order for the job to finish on time and it graphically shows the interaction between different activities. A graphic depiction of a CPM schedule may be called a *Pert chart, logic diagram,* or *CPM diagram.*

Planning the CPM Diagram

When scheduling using CPM diagrams, the processes of sequencing and establishing durations for activities are much the same as those for bar chart scheduling. The difference is in how the CPM diagram relates the sequences and durations of the various activities to each other. The Activity-on-Node (AON) style of diagramming is used by most builders today, in part because it is fairly easy to read and in part because most scheduling software uses this type of diagram. In an AON critical path diagram boxes called *nodes* are used to represent activities (work to be performed) and arrows are used to represent the relationships between activities. Each box contains a number identifying a single activity. For example, in Figure 5.2 the circle marked "29" represents the activity, "Heating and Framing Inspections."

FIGURE 5.1　The Bar Chart

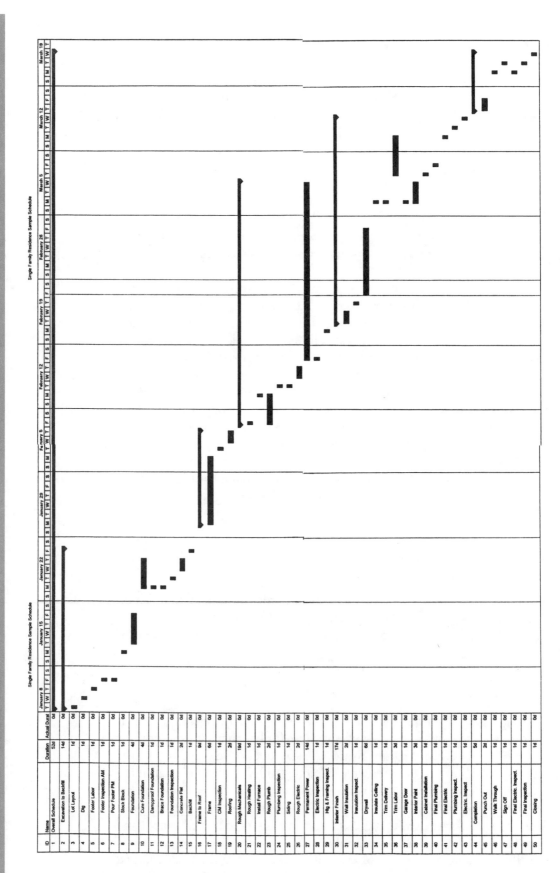

FIGURE 5.2 The CPM Diagram

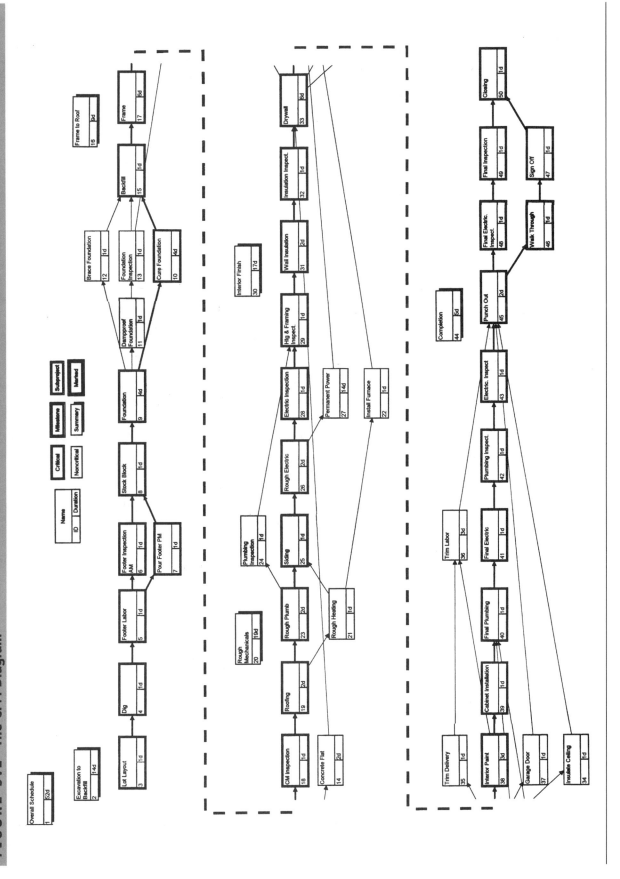

Scheduling with a CPM Diagram

Reading a CPM diagram is relatively simple. The basic rule states that all activities that lead into a particular activity must be completed before that activity can begin. For example, Figure 5.2 shows that the wood trim will be installed after the interior painting has been completed. You can easily follow the superintendent's scheduling approach. The first three activities depicted in the CPM diagram are creation of the overall schedule, excavation to backfill, and lot layout. Laying out the home can be considered the first construction activity. With the lot laid out, you can begin to dig. After the digging is complete, work on footings can begin, and so it continues. After dampproofing is complete, three tasks can begin:

- brace foundation
- foundation inspection
- cure foundation

Notice that after the completion of framing a construction manager inspection occurs, after which the roofing, HVAC rough-in, rough plumbing, and electrical rough-in may be done. Also notice that as these three jobs are completed, their inspections immediately follow.

Monitoring the CPM Diagram

During the monitoring phase of scheduling the superintendent can use the CPM diagram to the greatest advantage. One of the most effective monitoring techniques involves highlighting activities or parts of activities with a colored pen or highlighting marker. By doing so, you will be able to tell at a glance which activities are under way—and what is coming up next. If you are using one of the many computer programs available to do your scheduling, updating the schedule may be as simple as indicating on the schedule that the activity is complete, or 40 percent complete, or whatever the percentage may be. The computer automatically updates the schedule and makes adjustments as necessary. You can obtain an updated printout at any time. Material deliveries can be scheduled and will become easier to control. Finally, trade contractors can be lined up and schedule adjustments made more easily and accurately.

Scheduling Trade Contractors

The only accurate way to schedule trade contractors is to first consult with them individually and then negotiate a mutually agreeable duration for each activity. Resistance most often occurs when trade contractors are not offered an opportunity to participate in the scheduling process. Such participation should substantially improve trade contractors' willingness to establish realistic timetables and stick to them. Schedule control also can be made easier by taking the following steps:

- Hold a preconstruction scheduling meeting with all important staff, trade contractors, and suppliers involved on each major project or complex custom home. This meeting encourages coordination and cooperation among trade contractors and suppliers and reduces the likelihood of bottlenecks. (Even for routine production homes periodically holding a preconstruction scheduling meeting helps maintain positive communication and ensure coordination.)

- Avoid scheduling conflicting trade contractors on the same job, in the same place, at the same time.
- Have the job ready for the trade contractor on schedule; notify the trade contractor immediately if you learn of an unavoidable delay.
- Send written notices to trade contractors reminding them when they are scheduled for a particular job.
- Encourage cooperation among trade contractors. If they suggest useful and practical ideas that can help reduce costs or time while maintaining quality, accept the suggestions. Furthermore, recognize the trade contractor who suggested it.
- Base payment on compliance with the schedule. Trade contractors who adhere to the schedule should be paid on time. If the trade contractor delays the schedule, delay the contractor's check to correspond exactly with the number of days they delayed the job. Trade contractors will recognize the schedule as a priority when their checks are tied to it. (If you plan to implement this policy, however, be sure the trade contractors know about it up front. Also be sure this stipulation appears in your contracts with trade contractors, and be sure to apply the policy fairly and uniformly with all trade contractors.)
- Reward superior performance. If the trade contractor is responsible for a substantial savings as a result of superior performance, reward that performance with a bonus and a letter of commendation.
- Negotiate with trade contractors who are unaccustomed to working under a tight schedule. By negotiating from a position of knowledge and authority, you may achieve substantial time savings.

6

Construction Team Building and Trade Contractor Management

I n the past a construction company's reputation was often based on the skill of the craftspeople on its payroll. Skilled workers were a tremendous asset and construction companies often completed most of their work using hourly employees. For many years, however, the trend in residential construction has been a shift toward subcontracting more and more of the job, so that most building companies now subcontract most if not all of the work.

Hired Labor or Trade Contractor?

Many factors enter into a builder's decision to use hired labor or trade contractors for a particular construction job. The decision affects the superintendent in that the techniques and goals of managing trade contractors differ in certain important respects from those used in managing hired labor.

Advantages of Using Trade Contractors

The long-standing trend toward subcontracting reflects the fact that working with independent trade contractors carries a number of advantages for the builder.

Greater Flexibility. The demand for new homes fluctuates greatly, particularly in relation to mortgage interest rates. In times of high demand, building companies like to build as many homes as possible; when demand is low, they must be able to adapt to lower volumes of production. The more a building company relies on trade contractors the more quickly the builder can adapt to achieve this increased or decreased output.

Less Risk. By subcontracting work, building companies transfer some of the financial, employment, and management risks to the trade contractor. Since the trade contractors perform the work, they are responsible for the work. The builder does have ultimate responsibility, but only when the trade contractors refuse to honor their contractual commitments.

Less Capital Investment and Overhead. Trade contractors often furnish material, labor, tools, and specialized equipment. The building company can thus reduce its capital investment by not having to invest in costly equipment that may sit idle much of the time and by paying trade contractors only upon completion of actual work. In addition, because the builder can get by with fewer employees, tax and insurance costs are reduced.

Less Bookkeeping. Because the builder requires fewer employees and purchases fewer materials, less bookkeeping is usually required. You do not have to keep track of individual employee records.

Less Waste. When trade contractors supply their own materials more care seems to be given and damage and waste tend to be kept to a minimum.

Improved Quality. Trade contractors generally specialize and become skilled and efficient in one or two particular trades. This expertise makes the quality of work easier to control. If builders establish acceptable quality standards before work begins, they can usually ensure this quality by withholding payment until those standards are met.

Scheduling. Because trade contractors are independent, a builder can stipulate in the contract that certain schedules will be followed and that penalties will be imposed for noncompliance. One Chicago builder uses a "progressive network" to support the schedule. This builder's system requires each trade contractor to notify the next one when his or her job is complete so the next step can begin. For this type of system to work successfully, you will need good working relationships with all your trade contractors. Additionally, the notification requirement must be included in the trade contractor agreement.

Less Detailed Supervision. Superintendents usually are not required to act as job foremen or supervise trade contractors' day-to-day activities because trade contractors are responsible for their own work.

Disadvantages of Using Trade Contractors

Although the advantages of using trade contractors generally outweigh the disadvantages, some drawbacks are inherent when builders subcontract out most or all of the trade work.

More Coordination. The superintendent has a much greater coordination burden when working with trade contractors than with in-house employees. Each trade contractor represents an independent business. This fact may make it difficult to schedule trade contractors when you need them and to make sure they perform their work in a timely manner according to your company's quality standards. Executing those inevitable last-minute schedule changes can prove particularly difficult.

Unqualified Trade Contractors. Because of the low capital investment usually required, starting a subcontracting business is fairly easy. Many states do not require either a license or experience, resulting in trade contractors who may not be qualified to handle a job in terms of knowledge, experience, or adequate financing.

Supply and Demand. In a strong economic climate when demand for housing is high, a sufficient supply of trade contractors may not be available. The few trade contractors willing to take on more work may be overextended and unable to adequately fulfill all of their commitments.

Experienced Supervision Required. Superintendents must be qualified to distinguish high quality work from substandard work. They must be prepared to recognize problems immediately, suggest alternatives, and converse intelligently with those involved in the trades.

Trade Contractor Management

Finding qualified trade contractors can be a real challenge. There is a lot more to being a trade contractor than simply buying a pickup truck and a few tools. Hiring the wrong trade contractor can prove devastating to the builder. It is important to know what to look for.

Qualities of a Desirable Trade Contractor

The following are among the many qualities that, when combined, make a desirable trade contractor:

- financial stability
- quality workmanship
- loyalty to the builder
- cost-consciousness
- ability to stay on schedule
- dependability, honesty
- cooperation, teamwork
- strong "people skills"
- adequate, competent workforce
- prompt payment of bills
- prompt service on callbacks
- compliance with legal requirements (workers' compensation; license, where required)
- professionalism (business cards, easy to contact during reasonable hours, professional appearance)
- minimum material waste
- adequate insurance coverage
- adequate employee supervision
- clean, neat jobsite
- safe work practices, an effective safety program
- positive track record, time in the business
- fair prices, including the cost of all of the above

The Trade Contractor-Superintendent Relationship

A cooperative relationship between the trade contractor and superintendent is a vital aspect of a successful construction business. NAHB has sponsored hundreds of seminars nationwide to groups of builders and superintendents. In each seminar the presenter asks superintendents what they consider to be the greatest key to their success or the cause of most of the problems they face. Universally their response has been trade contractor management. Trade contractors either cause them their greatest problems or are the greatest contributor to their success. The relationship you have with your trade contractors can be the most important factor determining your success as a superintendent.

Trade contractors can literally make or break you. You should therefore encourage trade contractors to participate actively not only in the construction of your homes, but also in the whole construction process. Encourage them to suggest new, more efficient methods, products, materials, and techniques.

Keep in mind that while a trade contractor works for the building company in a relationship similar to that of an employee, the trade contractor is *not* the builder's employee. The relationship is one of one business to another business, and it is established and specified by means of an oral or (preferably) written contract. Because the builder's role is that of general contractor, the builder should stipulate the terms of this contract.

A trade contractor must make a fair profit in order to stay in business. Builders and their superintendents should be careful not to take advantage of trade contractors who are inexperienced or who have made an obvious mistake in their estimating. These trade contractors may provide work for a cheap price, but you can probably count on work that matches the cheap price. Disqualifying ridiculously low bids—or even giving a trade contractor the opportunity to back out gracefully—usually works out better than hiring an unqualified trade contractor.

Locating Trade Contractors

The best trade contractors are usually the busiest ones. Experience has shown that the best time to look for good trade contractors is when you do not need them. If you wait until you do need them, they are likely to be busy on someone else's job. Some builders do most of their recruiting in the winter months when things are a little slower. Trade contractors are often plentiful and you have a little more time to look for them and check them out. Help-wanted ads in local and neighboring newspapers are a useful source of potential trade contractors. During the winter months superintendents can sort through the responses, qualify the potential trade contractors, fill out contracts with them, and orient the new trade contractors to the builder's policies and procedures.

Superintendents should maintain a list of potential trade contractors by trade. As you drive around or when you encounter a potential trade contractor, take a moment to jot a note to yourself to investigate further. Keep a list of names of those trade contractors who have shown interest in working with your company, filled out an application, signed a subcontract agreement, submitted a worker's compensation certificate, or submitted a bid or price list. Prequalify trade contractors on the list before you need them so that when opportunity arises the trade contractors are already set up and ready to go to work.

Some production managers make the potential trade contractor list a regular part of their weekly production meeting. As they meet with superintendents they try to accomplish several things:

- review the need for trade contractors in each trade
- review the list maintained by each superintendent
- consider the status of each potential trade contractor (paperwork complete?)
- rank the potential trade contractors by observed work quality
- identify new trade contractors to assign to particular jobs and superintendents
- assign a superintendent responsibility for training the new trade contractor(s)
- look for areas of weakness in the trade contractor pool and identify future needs
- assign superintendents to recruit and qualify trade contractors for specific trades or areas of weakness

In order to find good trade contractors, you might contact your local home builders' association, other builders, suppliers, other trade contractors who work for you, inspectors, and other superintendents. Suppliers can be excellent sources of information about potential trade contractors. They generally know how much business the trade contractor is doing and how well they manage their business. Suppliers also usually know if a trade contractor is financially stable and will likely not recommend trade contractors who have a tough time paying their bills.

Inspectors eventually see every trade contractor's work and often are in an excellent position to compare the work of trade contractors fairly and objectively. They also should be considered sources of information about potential trade contractors.

While direct observation takes time, you also can get a pretty good idea of how a trade contractor performs by inspecting the contractor's other projects. Consider the following questions as you look at their work:

- Is the jobsite clean?
- Are the materials and the work quality good?
- How well does the contractor's crew use the materials provided?
- Is there a lot of waste?
- Does the trade contractor work well with other people?
- Does the job sit for long periods without much progress?
- Are sufficient workers on the job to complete it in a reasonable amount of time, or does the job appear to move too slowly?
- Are the workers professional in appearance and actions?
- Does the builder like working with the trade contractor?

Finally, check each potential trade contractor's current financial status with material suppliers (be sure to check with more than one) and credit bureaus. Financial standing changes quickly, so make sure the information you have is accurate and current.

Consider developing your own trade contractors. Sometimes you can help someone who wants to start his or her own subcontracting business to get started. There are substantial benefits in developing your own trade contractors. You can train them exactly the way you want, and new trade contractors usually have minimal overhead. However, they may also lack some of the necessary equipment. The trade contractors you develop tend to be extremely loyal. They take some time to develop, but in time they can be among the best resources you have.

Managing Trade Contractors

A superintendent must have the managerial ability to schedule, coordinate, and control all the trade contractors on several concurrent jobs so that work proceeds—as always—on schedule, within the established budget, and according to the quality specified. A superintendent also must be able to evaluate the managerial abilities of each trade contractor and determine if the trade contractor will be able to meet payrolls and overhead costs, pay suppliers, and still make a profit.

Building companies in even the best financial condition may find themselves in very difficult circumstances if trade contractors go broke in the middle of a job. Because the building industry is so dynamic and volatile, many builders' businesses have been seriously hurt by trade contractors who were poor businesspeople.

An adaptation of Pareto's law states that, according to the "80-20 rule," 80 percent of a builder's existing trade contractor base will perform a job to the builder's satisfaction or have the potential to do so with adequate training. This 80 percent is worthy of keeping and developing. The other 20 percent will cause 80 percent of your problems. They usually lack one or more of the ingredients necessary for success. How you deal with this 20 percent determines, to a large extent, the success of your construction operation. You are only as strong as your weakest link (trade contractor).

The Superintendent's Role

Your managerial role as a superintendent changes when trade contractors are used. Instead of being responsible for motivating and coordinating your own workforce you must direct and control highly independent trade contractors who, in turn, direct their own crews. You should therefore try to foster cooperation between the trade contractors themselves, particularly when their needs conflict, and promptly mediate an acceptable resolution when necessary.

You can make the supervisory task a smoother one by ensuring that jobs are ready when a trade contractor arrives. Check the jobsite in advance. Trade contractors often find themselves called in too early. Sometimes the superintendent assumes the job is ready when in fact the preceding trade contractor has not yet finished or the jobsite has not been cleaned up. The results are costly dead-end runs and extra trips. Make sure the preceding trade contractor is finished and out of the way, the needed materials are there, and the job is ready.

Give trade contractors as much lead time as possible to facilitate their scheduling and to avoid rushed, last-minute decisions. Also be sure to inspect the trade contractors' work in a timely manner so that any necessary changes can be made with minimal delay and expense.

Written Contracts

A wise person once said, "An oral contract is not worth the paper it's written on." It is true that oral agreements can be difficult to enforce when disagreements occur. Yet many builders and superintendents continue to operate on a promise and a handshake, often for the following reasons:

- lack of familiarity with contract provisions and law
- reluctance of trade contractors to be bound by written agreements
- lack of standard contract procedures in many building companies
- ability of many trade contractors to perform adequately without a contract

Written subcontract agreements are an absolute must. They are the primary means of communicating and coordinating expectations before the project begins. Consider them your opportunity to start the project off on the right foot. Without formal written contracts you leave yourself open for myriad problems, including but not limited to the following:

- problems with the Internal Revenue Service and with state agencies for trade contractor tax withholdings, workers' compensation withholdings, and accident liability issues
- problems over definition of who does what (scope of work)
- problems over noncompliance with the builder's safety program or with OSHA requirements
- problems over noncompliance with the schedule
- communication problems
- problems over definitions of quality and customer satisfaction (often after the project is complete)
- disagreements over terms of payment, retainage, and so forth
- conflicts between trade contractors
- disagreements over warranty and customer service

The greatest advantage of a written agreement is mutual and clear communication. Written contracts spell out a trade contractor's exact requirements and thus eliminate many areas of potential disagreement. They also can be used to align the trade contractors' work with the work promised in the sales contract with the homeowner. The Home Builder Press publication *Contracts with the Trades: Scope of Work Models for Home Builders,* by John Fredley and John Schaufelberger, provides sample subcontract agreements that can be used as a template from which to develop useful written contracts.

Subcontract agreements should include the following sections and provisions:

- a *general conditions* section that addresses issues common to all trade contractors, including responsibilities for workers' compensation, licensing, assignment of liabilities, and other matters
- a *trade-specific* section that addresses scope of work and details the materials specifications and acceptable methods of construction for each type of trade
- a *payment* section that addresses price and payment terms
- plans and specifications
- scope and quality of work
- scheduling requirements
- change order policies and procedures
- inspection policies and procedures
- warranty and customer service policies and procedures
- penalties for failure to meet contractual provisions
- payment provisions
- cleanup policies
- policies regarding use of facilities
- other policies, procedures, conditions, or provisions as needed for a specific project

Plans and Specifications. Specifications must define as clearly as possible exactly what is intended. Residential construction plans are sometimes superficial documents that do not begin to cover all of the important details. Vague specifications will confuse and frustrate trade contractors as well as the homeowners. Standards of construction should be established with detailed specifications that will meet the needs of the builder and eliminate guesswork. Any additions or changes to the plans and specifications should be properly authorized and submitted with accompanying change orders.

Scope of Work. Many if not most disagreements between builders and trade contractors stem from lack of definition of the scope of work. Written scopes of work should detail at what point the trade contractor's work begins, at what point it ends, and everything in between.

Quality of Work. The quality of work should be identified in detail. Every builder has different standards and methods for accomplishing the same work. With all of the different quality standards in home building, it is impossible for any trade contractor to know exactly what you expect. If you specify what you want in writing, your chances of getting it increase tremendously. Stipulate your expections up front when you are qualifying the trade contractor. Many builders use written performance standards for each trade. These standards normally become a part of the terms of the contract.

Scheduling

Scheduling trade contractors is normally the superintendent's responsibility (see chapter 5). To expedite construction, a work schedule must be coordinated with each trade contractor. The initial schedule, reflecting agreed-upon dates and deadlines, often is included as part of the contract documents to emphasize the trade contractor's commitment.

A superintendent's positive action, particularly at the early scheduling stage, can prevent many complaints. Matters likely to create controversy should be decided as far in advance as possible with the resulting decisions communicated to all trade contractors. The superintendent can then spend less time arguing and more time seeing to it that work gets done.

Change Orders

Improperly handled change orders are a frequent source of disagreement among trade contractors and between trade contractors and builders. Keeping accurate records on every requested change is extremely important. Every change request should be documented through a formal change order procedure that includes the following steps:

- request for change order
- change order cost estimate
- schedule update
- written approval signed by both parties
- change order deposit or method of payment
- completion of the requested change

The cost of the change order, as well as any increased costs related to rescheduling, should be passed on to the party responsible for the change.

Warranties and Customer Service

Customer service is one of the primary areas of homeowner complaints against builders. Most buyers are satisfied with their homes on settlement day but may become disappointed later when something goes wrong or they attempt to have a mistake corrected after move-in. The superintendent's frustration often is compounded when he or she is unable to motivate the appropriate trade contractor to handle customer service calls in a timely manner. You can eliminate a great deal of confusion, delay, and frustration by stipulating trade contractors' customer service and warranty responsibilities in their contracts, including the time limits permitted. More detailed information on working with the homeowner appears in chapter 7.

Payment Provisions

To the trade contractor the most important part of the subcontracting process is payday. Payment provisions should be clearly spelled out. Will the trade contractor be paid in phases, or must all work be completed before any payment is authorized? What constitutes completion of the work? Who inspects the work to make sure it is properly performed? Will there be a retainage?

Policies and Procedures. A builder's *policies* establish the philosophy of how the company will conduct business. A builder's *procedures* set forth the exact manner in which business will be conducted. Establishing policies and procedures for both running a business and completing a job is one of the first steps a building company must take to prevent problems, avoid disagreements, and eliminate confusion. Establishing policies and procedures is the role of the builder (or top management). Explaining and enforcing policies and procedures for trade contractors generally falls to the superintendent. Policies and procedures for trade contractors should encompass the following topics:

- scope of work to be performed
- lines of authority and channels of communication
- acceptance of materials on the job
- proper conduct on the job
- use of site facilities like power and equipment
- acceptable and unacceptable construction methods
- use of equipment, materials, and temporary utilities
- provision of adequate crews on the job
- compliance with schedules
- safety, accident prevention, accident management, and accident reporting procedures
- protection of other trade contractors' work
- acceptance of the work
- payment policies and procedures
- lien waiver and protection requirements
- liability insurance
- workers' compensation insurance
- cleanup procedures
- callback and warranty procedures
- special exceptions

Standard policies and procedures can be written up in a separate document that is referenced in the provisions of the contract along with the plans and specifications. Policies and procedures should be stated in a positive manner and should not be merely a collection of "thou shalt nots."

Training Trade Contractors

The first step in training trade contractors begins when you hire them. A number of documents and steps are involved when the superintendent sets up a trade contractor to work for a builder, including the following:

- Application: Have each trade contractor fill out a trade contractor application.
- Trade Contractor Agreement: The trade contractors should read and sign the trade contractor agreement for their particular trade. It is a good idea for the superintendent to review the agreement with the new trade contractor in order to highlight important items or draw attention to items included to prevent a recurrence of problems the builder has encountered in the past. Be sure to review the payment provisions. How should the trade contractor apply for pay-

ment? When are invoices due? When will they be paid? How will trade contractors and their crews receive payment?

- Plans: Ideally, give the trade contractor a set of plans for the house they will be working on (a typical set of plans will do for an orientation if specific plans are not available). Discuss what you expect from the trade contractor and how you expect the work for their particular trade to be performed.

- Standard Details: Give each trade contractor a copy of any standard details used by the company for their particular trade. For example, details for stairs, decks, and cornices should be given to framing trade contractors.

- Construction Standards: Give each trade contractor a copy of the construction standards or the section of the *Production Manual* that pertains to their trade. Emphasize any items that may be unique to your company.

- Quality Standards: Give each trade contractor a copy of the builder's quality checklist for their particular trade and go over it item by item. Make sure the trade contractor understands what each item means. Explain that a completed quality checklist is required from each trade contractor as a condition of payment.

- Change Orders: Explain the change order procedure for trade contractors and the need for signed change orders.

- Purchasing: Explain the process for obtaining additional materials when shortages occur and the variance purchase order procedure used by your company.

- Payment: Explain the payment procedures of your company.

- Safety: Explain the builder's safety requirements as they apply to the trade contractor's trade. Personal protective equipment (hard hats), fall protection (scaffolds with rails and harnesses), training requirements for trade contractor employees, and other safety precautions should be explained.

- Pricing: Obtain pricing information from the trade contractor. Be prepared! Know the range of prices paid for each subcontract trade and negotiate prices for standard items.

- Schedule: Discuss the importance of each trade contractor maintaining the agreed-upon schedule. Explain how the schedule is developed and updated. Discuss the impact on other trade contractors when one trade fails to comply with the schedule. If payment will be tied to schedule compliance, discuss the company's policies and procedures on this topic.

- Insurance: Obtain copies of the trade contractors' insurance and workers' compensation insurance certificates.

- Taxes: Review how tax forms (1099s) will work.

It normally takes from two to three hours to properly orient a new trade contractor to your company's policies and procedures. Numerous topics must be covered, and allowing time for discussion is an important part of the orientation. At this point, both of you are normally anxious to work with each other. Take this opportunity to train the trade contractors and to make sure all of their questions or concerns have been addressed. This is the best time to make sure you are both working under the same assumptions or rules.

When new trade contractors start on their first job for you, be there when they arrive. You can use the opportunity to get them oriented to the site and to discuss any pertinent topics or issues unique to that particular job. This is a good time to reemphasize the importance of the schedule and standards of construction. Show the new trade contractor where to put their trash and how you expect them to clean up.

Plan to set aside a considerable amount of time to work with the new trade contractor on their first job. Many superintendents spend as much as 50 percent of their time orienting and working with new trade contractors. If you are on the site when they arrive and stay long enough to make sure they get a good start, then you can go work on your other jobs for a while. Depending on the trade and the contractor, you may want to return to the job a little later to answer any additional questions they may have and to make sure things are going well.

If the new trade contractors will be working on the jobsite longer than one day, make sure you are on the jobsite when they quit the first day. This allows you another opportunity to answer questions and see how things are going. If your company furnished the materials check to make sure they are being used correctly and to see if any additional materials are needed. Before the new trade contractor finishes, check the work thoroughly. Help the trade contractor fill out the quality checklist. Make sure the jobsite is clean and the work is complete. Often there will be things the new trade contractor can do to increase their productivity and improve their profit margins. Show them ways they can improve performance. Do what you can to make sure the first experience with your company was a positive one for the trade contractor. The first job is your best opportunity to establish a positive, long-term relationship based on the standards and expectations to which you train the new contractor.

The Dangers of Familiarity versus Single Source Suppliers

Although building companies often find it easier to deal with the same trade contractors on numerous jobs, the builder and superintendent must determine the relative importance of price, service, quality, and management complexity as they relate to subcontracting. One of the basic tenets of the TQM philosophy is the concept of a *single source supplier*. Working with a single source supplier has several advantages for a builder, including the following:

- firm price on a regular basis makes estimating easier
- ease and efficiency of training
- consistent quality standards and methods of construction
- familiarity and compliance with builder's policies and procedures
- loyalty generated between the builder and trade contractor
- ease of working with familiar trade contractors
- ease of scheduling

There can be drawbacks to using the same trade contractors over and over, however. Using the same trade contractor on every job may result in increased cost if the trade contractor is not competitive. Building companies should avoid increased costs merely to simplify trade contractor management.

If a building company does decide to use the same trade contractors on a continuing basis, superintendents must continue to check regularly to ensure the trade contractors remain competitive and are maintaining the builder's standards of quality. You also may want to occasionally check the financial status of familiar trade contractors, particularly in an environment of changing economic winds. When a superintendent, builder, and trade contractors become too friendly, the danger exists that each side may begin to take the other for granted, potentially leading to problems. To counter these problems, some building companies routinely request competitive bids from several trade contractors. This practice ensures a more competitive price, provides better control, and maintains a higher level of service.

Partnering With Trade Contractors

In recent years builders have learned the importance of partnering with trade contractors. Some builders have been able to reduce the cost of construction substantially by cooperating with all of their trade contractors in an overall cost-reduction effort. Superintendents can facilitate builders' partnering with trade contractors by taking the following steps:

- maintaining a strong, qualified list of current and potential trade contractors
- providing adequate orientation and training to new trade contractors
- enforcing the builder's policies and procedures uniformly and fairly
- mediating disputes or disagreements that arise among trade contractors or between trade contractors and suppliers when necessary (but without micromanaging crews on the jobsite)
- respecting the expertise of the various trades
- listening to suggestions from trade contractors and suppliers, organizing and implementing those that prove useful to the builder or the project
- recognizing the contributions trade contractors and suppliers make to a successful project
- coordinating the communications and efforts of various independent contractors to produce a positive, trusting work environment

Measure Performance

What is your current performance? How long does it take to build your homes? What are your cost variances for each activity or cost category? What is the quality of the work performed? Are there quality problems and rework for each different trade? How is your productivity? Are you spending too many hours to perform certain tasks? How is your customer service? How many outstanding service work orders do you have? On average how long does it take to complete a service work order?

Partnering for Greater Efficiency

Trade contractors can help each other save substantial sums of money and reduce costs by cooperating together while increasing individual profits and the profitability of the builder. For example, it is extremely difficult to insulate properly behind an installed tub or shower enclosure. The insulation may be left out entirely or be stuffed into the top portion of the wall—which means it does not fill the stud wall cavity. Yet wall insulation is not normally installed until after the plumbing, electrical, and HVAC are complete.

One solution is to have the insulation contractor come out and install batt insulation in the wall after framing but before the tubs and shower enclosures are put in by the plumber. This works, but the extra trip by the insulation contractor is costly and

the interruption of framing or plumbing work may create friction.

By partnering with trade contractors you can get them to cooperate with each other, eliminating the problem entirely. Some builders would supply the framer with a role of batt insulation and vapor barrier material that they can install behind the tub and shower enclosures when they finish framing. Other builders would provide the materials to the plumber, who would insulate the wall before installing the tubs and shower enclosures. This solution eliminates an inconvenient trip for the insulator. The framer or plumber generally will not mind doing the extra work if they are compensated for their time, the work is minor, and the materials are supplied by the builder. The cost is very minimal.

Be Open to Suggestions

Listen carefully to your trade contractors. Encourage them to suggest better ways of building your homes. Do not be afraid of criticism. Do not be defensive or paranoid. Find out what causes problems for your trade contractors. Sit down with your trade contractors and ask what you can do to help them improve their work and help them make more money. Write down their responses. Consider holding a masterminding dinner with trade contractors to promote suggestions.

Organize the Ideas

Organize all of the ideas and prioritize them.

Implement the Ideas

Develop a pilot program to implement the ideas. Measure the difference in terms of cost, time, and quality. Measure customer acceptance and satisfaction. If the results are positive, implement the ideas that work company-wide and modify or reject the other ideas.

Keys to Successful Partnering with Trade Contractors

Superintendents can do a number of things to foster positive relationships with trade contractors, including the following:

- Be ready when they arrive. This is important with veteran as well as new trade contractors. Few things are more frustrating to a trade contractor than arriving on a jobsite ready to work only to find that the preceding trade contractor is still there or that the job is a mess. Make sure the preceding trades are finished. Be on the jobsite early so the trade contractor does not have to sit around waiting for instructions. Have the necessary material stocked and readily accessible. Make sure each trade contractor cleans up when their work is done.
- Check the work of new trade contractors early and strictly. Not only is careful checking important as part of the new contractor's orientation, it also sends a message to the other trade contractors that you expect the new contractors to perform at the same high level. Both new and veteran contractors will respect you for it.
- Keep the Jobsite Clean. Good housekeeping pays! Insist that trade contractors clean up the job every day. Ten to fifteen minutes of cleaning at the end of the day will save them 30 minutes of prep work the next day. A clean jobsite will increase customer satisfaction and referrals. It is also much safer and will decrease potential liability and injuries.
- Monitor progress and provide feedback. Let the trade contractors know how they are doing.
- Pay on time all the time. Nothing is more important to a trade contractor than getting paid on time. Nothing you can do is more important or will bring a greater payback. Process pay requests and approve appropriate invoices promptly. Monitor the pay situation to make sure trade contractors receive their paychecks on time. If delays occur, find out why and take steps to correct the bottlenecks. When there are problems getting the trade contractors paid let them know and try to mitigate the impact.

- Keep accurate records. When working with trade contractors it is important to keep accurate records. Record how often they failed to show up or finish on time, the time it takes them to complete their work, the number of service work orders, the time it took them to complete each service work order, and their performance with regard to safety compliance, among other things. These records will be important in partnering, improving performance, and when it comes time to renegotiate contracts.
- Don't cede control over the project to the trade contractors. You are the manager. Maintain control.
- Above all, *communicate!* Communication is probably the biggest key to working successfully with trade contractors and the key to communication is to listen, listen, listen. The trade contractors often know more about the job than you do.

Hired Labor

When a building company chooses to perform the majority of its field work with its own hired crews rather than employing trade contractors, the superintendent's job changes markedly. As a superintendent working with hired labor you will be less concerned with coordinating various crafts and more concerned with planning work to keep workers productively busy. Besides having increased hiring responsibilities you will have increased training responsibilities.

Although the hiring principles and practices discussed in this book's introductory chapter were written for the builder hiring a superintendent, they also can be applied by the superintendent hiring members of a crew. Some additional comments are in order, however. One of the most important steps in hiring is the assessment of organizational needs. Most builders and superintendents face hiring problems when they fail to define their needs clearly at the outset. Planning is therefore essential to effective hiring.

Some superintendents fall into the trap of looking at their wants rather than their needs. These superintendents usually end up hiring people similar to them, believing that people like themselves are most likely to be successful. While having a protégé can certainly satisfy a superintendent's ego, hiring workers with substantially different, yet complementary, backgrounds will normally offer greater rewards in the long run. With so many different and diverse skills required in the construction industry, it only makes sense to hire someone who will add to the building organization rather than fit an existing mold. For example, if personnel are currently available to frame houses but lack the ability to install finish work, an employee who can both frame and install finish work would be an important addition. On the other hand, if a building company has workers for framing and layout who are not as productive as they could be, you should look for a productive worker with qualities that will motivate others.

In assessing company needs, superintendents should do the following:

- Seek opinions from the people within the company who will be working with the new employees.
- Talk to trade contractors about any needs they see.
- Look at the competition. What type of workers make up their organizations?
- Talk to other people who know the building industry in general and your company in particular.

Prepare Job Descriptions. Once your company's needs have been clearly identified, the builder and superintendent can work together to prepare written job descriptions. The job descriptions can serve as checklists for assessing applicant qualifications and comparing them with the needs of the organization. These job descriptions should be comprehensive and provide information about the following topics:

- all functions to be performed
- job responsibilities
- resources available
- the extent of authority of the position
- relationships with other members of the organization
- means of evaluating performance

Job descriptions may already be available for many positions. However, if company needs require a new or revised position, a new job description should be prepared. In addition to technical or mechanical skills, emphasize the more general attributes and abilities the job may require, such as the following:

- knowledge of construction
- common sense
- initiative to get the job done and take necessary action
- ability to confront problems and deal with potentially unpleasant situations
- oral and written communication skills
- leadership skills
- organizational skills
- flexibility
- listening skills (good people skills)
- ability to adhere to procedures and monitor processes
- ability to control stress and maintain stable performance levels
- ability to learn and apply new information
- motivation for work
- necessary mathematical skills
- an understanding of spatial relationships, particularly in visualizing shapes from drawings
- mechanical reasoning skills
- ability to read and understand blueprints and specifications

A complete and accurate job description is one of the superintendent's best hiring guides and can also be helpful when evaluating individuals for promotions and pay increases.

Assess Current Employees. Experts in personnel management constantly stress the importance of promoting an organization's present employees whenever possible. This preference over outside recruiting is understandable given the riskiness of hiring new people. Promoting current employees minimizes this risk, since the odds of objectively evaluating the skills, abilities, and knowledge of current employees is greatly enhanced. In addition, a work environment that offers employees the opportunity to grow professionally is a healthy one. In such an environment performance tends to improve and turnover declines. Many large organizations have a policy that one or more people should be in training for each manager's job at all times. This policy provides continuity to the company and reduces the tendency toward "management by crisis."

Current employees should be evaluated according to the following criteria:

- Does the employee have a positive attitude toward work and himself or herself?
- Is the employee technically competent to do the job?
- Does the employee understand the complexity of the business and the role of his or her job in the company's profitable operation?
- Does the employee have a good record of attendance?
- Does the employee look for ways to become more productive—and make other employees and the job more productive?
- Is the employee loyal to the company, to company management, and to the building industry?
- Does the employee get along with others? Can the employee motivate others to maximum performance?
- Does the employee have the potential to continue growing with the company, or is the employee at or near a "level of incompetence"?

The extent to which current employees meet these criteria will determine how much recruiting will be necessary. If the requirements are met in their entirety, new prospects should not be needed. However, if current employees fail to meet the criteria, you may want to recruit enough outside individuals to conduct a comparison of all available candidates. If current employees simply are not qualified, more extensive recruitment will be necessary. Current employees who apply for jobs and are subsequently turned down should be informed of the reasons for their deficiency and the improvements required. If this task is done with tact and in a positive manner, it can motivate employees to greater performance.

While promoting individual workers within the building company can prove beneficial, new employees will inevitably need to be hired from outside the organization on occasion. It is essential that these new employees have potential for growth. Planning for future needs is just as critical as planning for current needs. Too often a building company will wait until the last minute or until a need is critical before giving much thought to hiring new employees; consequently, the company hires the first person available who is reasonably qualified, ignoring the long-term impact of this hurried decision.

Training Hired Personnel

Once the right person has been hired for the job, you will need to turn your attention to training the new employee. Because the superintendent is responsible for training the field staff in new, more effective methods of construction, training and education are important keys to your success. They can make the difference between a growing, dynamic building company and a stagnating, declining one.

Training provides benefits not only to the new employees but also to the building company, as follows:

- Training improves job performance.
- Employees take pride in a company that devotes the time and money necessary to provide training.
- Training is a two-way process. The "I care" attitude training creates often improves communication, which in turn encourages suggestions from employees that may improve performance.
- Training in safety procedures and methods reduces hazards and improves a company's safety record.

■ Training employees in the best and most economical ways to perform certain jobs gives the company a competitive advantage.

Induction. Induction is the first important step in training a new employee. Effective induction procedures for each new employee can establish a positive attitude toward the new work environment for some time to come. During this initial period, the worker either confirms or changes his or her expectations of the building company.

Almost every job in a building company can be done many different ways, many of which are correct. New employees are often apprehensive about the way they have been taught and will most likely take several days to adjust to their new work environment. One means of easing this initially awkward period is to assign a particularly personable employee to work closely with the new employee for a few days to answer questions and help the new person adjust.

Orientation. Orientation is the next step, a time when the new employee becomes comfortable with the work environment. New employees are commonly concerned with the following matters during the orientation period:

■ relationships with others in the organization
■ tools and equipment policies
■ payroll and timecard policies
■ quantity and quality standards
■ rules and regulations for conduct on the job
■ involvement in apprenticeship programs (if applicable)
■ advancement

Most of these concerns can and should be handled in a company policies and procedures manual. Unfortunately, many smaller building companies—and even some larger ones–have not developed these manuals. NAHB has developed a personnel manual template entitled *Personnel Handbook for Small Volume Builders.* The template comes on a computer disk that makes it possible for even small companies to quickly adapt it and develop their own employee policy and procedures manual. Superintendents and builders can work together to create a company manual that helps each new employee answer six basic questions:

1. What is my present position?
2. What are my responsibilities?
3. What are my rights?
4. What limitations do I have?
5. What are the policies of the company in different areas?
6. Where do I go when I have a problem?

Training Methods. Several methods are available to improve new and current employees' knowledge, skills, and attitudes. Among them are the following:

■ apprenticeship programs
■ on-the-job training programs
■ planned work activities
■ individual instruction

Apprenticeship programs lasting from one to five years are the traditional means of training tradespeople. By rotating from one operation to another and receiving related technical instruction, apprentices acquire a broad range of skills, master the application of skills already learned with speed and accuracy, and develop independent judgment. This training method enables these individuals to be productive throughout the training period. Studies have shown that workers completing an apprenticeship are more highly trained, work more steadily, learn new jobs faster, and are more likely to be supervisors than workers trained in other ways. In the United States apprenticeship programs must be registered with the State Apprenticeship Council or the Department of Labor Bureau of Apprenticeship and Training.

On-the-job training programs have proven successful for many building companies. Similar to traditional apprenticeship programs in many respects, on-the-job training rotates employees from job to job, giving them a broad background in various aspects of the building business. Experienced workers often are used to teach employees the necessary on-the-job skills.

Planned work activities can be an effective training method for specialized jobs. For example, carpenters and supervisors from an experienced crew can be asked to train workers in the more effective techniques used in wood-frame house construction. This type of training can eliminate much of the expensive trial-and-error learning that is too common in the building industry.

Individual instruction may be necessary on occasion to teach a worker a particular task. The training steps offered in Figure 6.1 may prove helpful if you find yourself providing individual instruction.

FIGURE 6.1 How to Instruct

- Prepare the trainee.
- Put the trainee at ease.
- Introduce yourself to the trainee.
- Get acquainted.
- Show the trainee the whole job.
- Eliminate fear.
- Assure the trainee that he or she can learn the job.
- Stress the importance of following and completing every instruction.
- Focus attention.
- Find out what the trainee already knows about the job.
- Select the one idea that offers the best starting point.
- Maintain the trainee's interest in the job.
- Explain what the trainee will gain (skill, knowledge, and so forth).
- Spell out any ground rules.
- Explain the need to ask questions.
- Describe the format of the training.
- Present the key points of the lesson.
- Demonstrate and explain.
- Have the trainee do the job.
- Ask questions and encourage the trainee to ask questions.
- Review procedures and routines.

7

Working with the Homeowner

As a construction superintendent you will spend the majority of your time on the jobsite supervising the construction process. However, you also play an important role in customer satisfaction, in helping buyers achieve the "American Dream" of homeownership. It is essential that you understand homeowners and their needs in order to meet those needs effectively and profitably.

Understanding Homeowners

While it may seem a little odd to some people, most homeowners consider their homes an outward extension of their hopes and dreams. To a buyer, anything short of his or her mental picture of the perfect home can be devastating. The buyer most likely came to the custom home builder or breezed into the sales office loaded with ideas fueled by magazines. These publications were likely full of glossy photo spreads of million-dollar homes and articles telling homeowners how to deal with builders and get the most for their money. Some of the buyer's ideas may have merit while others may not and the buyer may refuse to listen to reason or discuss a point rationally. Recognizing that a new home is usually the largest single purchase most people ever make may help you to understand what might otherwise be viewed as strange or erratic behavior on the part of the buyer. Remember to look at a given situation through buyers' eyes to better understand their motives and to avoid potential problems.

The Superintendent's Role in Homeowner Relations

Your main goal as a superintendent is to maximize the company's profit in the long run. Earlier chapters have discussed how bringing the job in on time, within the budget, and according to the quality required is essential to achieving this goal. While accomplishing these three objectives is a key element in satisfying the homeowner, there is more to it than that. A superintendent who does all of the above yet who fails to deal effectively with homeowners is not maximizing profits in the long run.

Why? It has been said that the greatest asset a construction company can have is a good reputation. Happy, satisfied buyers will spread the word. Customer satisfaction is largely a matter of the following three factors:

- the quality of the work
- the way customers are treated
- time

The third factor—time—is too-often overlooked. You may have found that the buyers of houses that fell behind schedule were more difficult to please in terms of workmanship. Likewise, for houses that were ahead of schedule, buyers may have been more willing to overlook minor deficiencies. The easiest way to cut down the number of items on your punch list and increase the satisfaction level of the customer is to stick with your schedule. Reduce your schedule and you will likely reduce the length of your punch list and increase customer satisfaction tremendously.

Keep in mind that many buyers feel that they are buying not only a new home but also a large dose of personal attention. Regardless of the size of your company and the number of houses you have under construction at any given time, homeowners are concerned with only one home—theirs. They expect the construction of their home to go without a hitch. They understand that you may have trouble with a few other projects, but not with theirs. To them, their home comes first.

Direct Involvement with Homeowners

The amount of direct contact a superintendent has with homeowners varies greatly from one company to another. In many larger companies the superintendent has little direct interaction with clients. The customer comes to a sales office and works with a salesperson to select their home and lot. The superintendent builds the home from a set of standard plans and the homeowner moves in with little direct interaction with the superintendent. Homeowners may visit the jobsite from time to time during construction or they may not see it at all until the home is complete. In companies that

---

The Customer Is Still the Customer

One of America's most successful home builders has a favorite saying that he promotes to maintain customer-awareness in his company:

"There are three rules to customer satisfaction.

1. The customer is always right.
2. When in doubt, refer to rule number one.
3. Even when the customer is wrong, he or she is still the customer."

Following this philosophy has worked well for this builder's company over a long period of time. While building over 650 homes a year, the company has a referral rate over 50 percent. More than half of the homes the company sells each year result directly from referrals by previous satisfied customers. Since they began concentrating on satisfying the customer, the company's advertising budget has been cut almost in half.

For a small-volume custom home builder these numbers may not seem too impressive, because almost all custom home sales come from referrals. But for a large- volume builder a referral rate of over 50 percent is very admirable. This company also consistently boasts a customer satisfaction rate between 93 and 96 percent. On one survey administered to each customer over 90 percent of it's customers indicated that they would recommend the company to their family or a close friend. This company works hard to build homes in a way that meets or exceeds the homeowners' expectations.

use this kind of structure, putting the homeowner in direct contact with the superintendent may create confusion and therefore may be undesirable.

Most custom home builders conduct business in a totally different manner. Often the builder works from plans provided by the homeowners or prepared by an architect. The homeowners are an integral part of the construction process. They often make decisions and select products for the home throughout the construction process. Interaction between the superintendent and the customer is extensive, and they may even meet daily during crucial periods.

Each company should set policies concerning the amount and type of interaction between the buyer and the superintendent. The lines of communication should be well established, communicated to all parties, and well maintained. When lines of communication break down, gentle redirection to reestablish communication according to company policies is in order.

Post-Sale Letdown

Every superintendent should keep in mind that a certain degree of "buyer's remorse," or post-sale letdown, is common after any purchase, particularly one as large as a new home. Buyers suffering from post-sale letdown may feel anxious that perhaps they have made a bad deal or have committed themselves to spending too much money. While handling post-sale letdown may seem like a job for the sales staff, superintendents must do their part to reduce the symptoms. Simple, reassuring statements such as, "You're going to love living in this house," "This is a beautiful home," or "This is a great neighborhood" go a long way and can help buyers feel confident that perhaps their decision was a good one after all.

Policies and Procedures

Customer relations policies and procedures normally depend largely on the contractual relationship between the builder and the homeowner. For example, a job performed under a lump-sum or fixed-price contract puts you in more of an "arm's-length" relationship with the buyer than does a cost-plus contracted job. In the latter relationship the buyer deserves to know not only the cost of different items of work but also why certain procedures are used.

Working under a cost-plus arrangement opens your accounting books to the customer. One of the most important factors in a cost-plus arrangement is determining what is part of cost and what is part of the plus side of the equation. Making this determination is sometimes difficult, however, because a number of items can be caught somewhere in the middle. Do project overhead costs such as security, temporary power, supervision, insurance, come under the part of the costs that the customer agrees to pay—or do they come out of the contractor's pocket? Is company overhead included in the cost or does it come out of the builder's profit? In cost-plus arrangements these things need to be clearly spelled out before construction begins.

Another primary factor to consider is whether an architect is involved and, if so, to what degree. If an architect is involved, he or she can be enlisted to help answer a buyer's questions and explain construction procedures. In many cases the architect will be the one to authorize payments to the builder as construction progresses.

As noted above, custom builders' policies and procedures with regard to lines of communication, site visits, and so forth should be carefully spelled out. Many custom builders make it a standard policy that the superintendent will communicate with the homeowners at least on a weekly basis to bring them up to speed on the progress of their home.

Increasing Buyer Understanding

Builders' policies should be designed to support the buyers' understanding of exactly what they are buying. One important reason for the preconstruction conference with the buyer (discussed in chapter 2) is the opportunity to establish the homeowners' expectations.

Buyers often have difficulty managing the leap from construction drawings to an actual house. By showing comparable room sizes, layouts, and details, you can enhance the buyer's awareness of the true finished product. Providing this attention within reasonable limits can lay a foundation for smooth relations with buyers. Perhaps even more important, however, are the homeowners' expectations for the quality of finishes. If the buyer was shown a model or other sample home, you must ensure that the buyer's home meets or exceeds the model in every respect.

The Importance of Contracts. When dealing with homeowners, whether in relation to a custom home, semi-custom home, or completely speculative home, there is no substitute for a thorough, easy-to-understand contract. Buyers naturally fear "snake oil deals" and contracts with the proverbial "small print." A customer-oriented builder or superintendent will take to time to make certain the buyer understands everything in an agreement.

One of the purposes of the preconstruction conference is to review the contract requirements in detail. It has been the experience of many builders that you have to constantly remind homeowners of their obligations under the contract, especially when building custom and semi-custom homes. As a practical matter, homeowners tend to forget what is required of them and the time frame in which things must be decided upon and completed. Many builders have found it helpful to include homeowner responsibilities as activities on the project schedule and give homeowners weekly reminders of their responsibilities.

Buyer X

Some buyers, unfortunately, remain unsatisfied even when the home meets or exceeds the quality of the model. For example, Buyer X had sold his first home and was moving up to his dream home. He had seen examples of the builder's finished product and was satisfied. One night, after his house was completed, Buyer X held a spotlight up against the walls and marked each and every imperfection. In addition, he complained that the tops of the door casing in one of the closets had only one coat of paint. While the builder agreed to try to fix the drywall imperfections, he explained to Buyer X that many of the imperfections would not have been noticeable under ordinary lighting conditions. To prove the point, the builder took Buyer X and his spotlight back to the home X had lived in for years. The spotlight test showed large imperfections that the homeowner had never noticed while living in the home. A look at the tops of the door casings in the closets of the prior home revealed not only the absence of a second coat of paint, but the absence of any paint at all.

Handling extreme cases like this one with tact and understanding can prove difficult at best, but the dividends can be great. You must always remember that you are a professional and must conduct your superintendent duties in a professional manner at all times. By doing so you can turn a potentially bad situation into one that enhances your company's long-term reputation.

Buyer-Requested Changes. An important area to discuss with every buyer—and include in every contract—is the policy regarding changes to the work. If an architect is involved, the architect's understanding of change orders should smooth the process considerably. In any event, two things must be made clear regarding changes:

- One person will authorize changes on behalf of the company.
- All changes will be written and will include an exact description of the change, the difference in the sales price, and changes to the schedule and method of payment, if any.

Use of a change order form that organizes the information needed in accordance with the builder's policies and procedures can be helpful in managing change orders (see Figure 7.1). The role of the superintendent should be carefully defined with regard to change orders. Does the superintendent have the authority to approve change orders? Can the superintendent price out change orders? How do you mark up change orders for overhead and profit? What does the superintendent do in case of an emergency? At bare minimum the superintendent should make it a point to stay completely up to date on the status of all change orders. Often the superintendent is the initiator of change orders and is directly involved with the customer in making sure they are completed.

Homeowner Visits and Company Contacts. Some construction companies have policies stipulating when homeowners may visit the jobsite. Some companies also pair each homeowner with a contact person within the company to arrange site

---

FIGURE 7.1 Change Order Form

Buyer: _____   Lot number: _____

Contract dated: _____   Plan: _____

Item Number	Description of Change	Cost	Schedule Adjustment
_____	_____	_____	_____
_____	_____	_____	_____
_____	_____	_____	_____
_____	_____	_____	_____
_____	_____	_____	_____
_____	_____	_____	_____
_____	_____	_____	_____
_____	_____	_____	_____
_____	_____	_____	_____

Totals: $ _____   _____Days

The changes listed above and the corresponding costs and adjustments in the construction schedule have been requested by you, the purchaser. By signing this change request you agree to pay for indicated changes and acknowledge that the construction schedule and estimated delivery date for the referenced home are revised accordingly. Change requests will be incorporated into the home only after they have been approved and signed by the builder.

Approved: _____   Purchaser: _____

Builder: _____   Purchaser: _____

Date: _____   Date: _____

Source: Carol Smith and William Young, *Customer Service for Home Builders* (Washington, DC: Home Builder Press, National Association of Home Builders, 1990).

Change Order Enforcement

Consistent enforcement of the builder's change order policy is especially important on cost-plus homes and in custom home building. In a court case in Maryland, a builder failed to keep the buyer up-to-date on cost overruns on a cost-plus job. The builder subsequently had to absorb these overruns because he failed to keep up with them in an orderly fashion and keep the buyer informed. The court ruled that cost-plus jobs are different from ordinary fixed-price contracts in that it is duty of the builder to act in the buyer's best interest. Therefore, the builder was obliged to keep the owner advised regarding the cost overruns so that the buyer could make appropriate decisions and adjustments.

visits and answer any questions that may arise. With most homeowners, such procedures are unnecessary. However, a homeowner will occasionally come along who is on the jobsite constantly, asking questions of everyone. Such people usually are just showing a keen interest and curiosity and do not realize that their presence is disruptive to the normal job rhythm. In certain cases, you will have to make buyers aware—in a tactful manner—that the builder is not trying to hide anything or get away with shoddy workmanship or substandard materials. Explain that in order for the job to flow smoothly, jobsite visits must be at preappointed times, and questions must be channeled through one person. This contact person may be the superintendent or someone on the sales staff.

For safety and liability reasons, you may request homeowners to limit their jobsite visits to prescribed times. It is important that homeowners remain sensitive to the work going on at the time of the visit. Their safety is your responsibility when they are on the jobsite. This issue should be covered in writing in the contract and emphasized during the preconstruction conference.

Despite your best efforts, you never really know when homeowners (or their friends or relatives) will show up on your jobsite. Often they will visit the job after hours or on weekends. To prepare for this eventuality, take the following steps:

- Spell out in writing the builder's policy with regard to jobsite visits. Include a statement concerning homeowners' children. Jobsites are inherently unsafe for small children.
- Make sure the jobsite is kept clean and neat. Every night and especially on weekends, the job should be cleaned and the material stacked (banded if possible) and properly stored. Excess material, waste, and especially trash (including lunch sacks, beverage bottles or cans, and cups) should be put in the trash dumpster or pile.
- Check the jobsite for safety. The jobsite should always be in a safe working condition, but especially when a customer or visitor is on the site. If you are performing dangerous operations such as erecting trusses or putting down subflooring, you may want to ask the customer to come at a safer time or at least put on a hard hat and other safety and personal protective equipment.
- Train the trade contractors regarding how you want them to deal with visiting homeowners. How should trade contractors respond to homeowners' questions or concerns? How should the trade contractors deal with change requests or questions about costs and prices initiated by the customer? This training needs to take place as part of the trade contractor orientation long before the homeowner arrives on the jobsite.

One way to effectively manage homeowners' jobsite visits is to arrange specific times to meet with the homeowners on site. Some superintendents meet weekly with homeowners at a specified time. They go over what has been done and make sure that any customer concerns have been addressed. There are certain times when a jobsite visit with the homeowner can be beneficial to the superintendent. Most superintendents conduct a framing walk-through. When framing is complete and the rough mechanical has been installed, it is a good time to make sure the customer feels comfortable with the home. You can walk through the home and point out where outlets will be located, what direction doors will swing, and where cabinets will be installed. The framing walk-through also is a good time to discuss light fixtures and furniture placement. (It is better to face furniture problems before the drywall is up and the house is painted, when it is too late to do anything about it.) Make sure the home is in good shape when the homeowner arrives. Make sure that all customer-requested changes are up-to-date and complete. Of course if additional changes are necessary, written change orders can be completed and signed during or after the walk-through. An additional walk-through is sometimes helpful after the drywall is installed or the painting has been completed.

Walk-throughs with the homeowners foster confidence and can be a great way to demonstrate the high quality of a home's construction. The customers recognize that you care about their satisfaction and concerns, and that you acknowledge their natural interest in their new home. Furthermore, talking through homeowners' concerns as part of one or more walk-throughs may alleviate or eliminate conflicts later, when they would be more difficult to resolve.

Communication

Good communication is probably the most important key to positive customer relationships. In most cases homeowners will be much more comfortable if you maintain frequent communication with them. People require different levels of communication, however. Some want to be left entirely out of the picture. "Call me when it's done," is their philosophy. Others need to feel deeply involved. Such people may require a daily phone call and a weekly site visit. It is important to determine the level of homeowner involvement and desired level of communication. The ideal level of communication may change from time to time. When the house is being framed or is in the finishing stages the homeowner may want more communication than when the footings or drywall are being done. At critical stages during the construction process it may be wise to increase the frequency of communication. In all cases, being proactive in communicating with the homeowner is much better than passively waiting to address customer complaints.

Fixing Homeowners' Concerns

Veteran superintendents know that homeowner concerns need to be addressed immediately. Even if a problem would have been handled in due course anyway, sooner is better. Many builders have their framer do a framing punch out after the HVAC, plumbing, and electrical activities have been completed as a matter of standard practice. Any framing that has been cut, damaged, or removed can be corrected and additional backing installed before drywall. Most homeowner concerns are handled naturally at that time. However, if a homeowner points out a concern to you during the framing walk-through and it is not corrected quickly, remember: the homeowner will likely focus on the problem every time they think about the house. If the problem persists, it may become a nagging sore spot. Most superintendents have found that it is better to fix the problem immediately rather than wait for it to be corrected naturally.

Promptly correcting problems that homeowners point out will accomplish several things:

- The homeowners will know you care about their concerns.
- The homeowners and the superintendent can both cross them off the "to do" list and stop worrying about them.
- The homeowners will not have to worry that you may forget about a problem or cover it up before they have had a chance to check it out.
- The customer will feel more confidence in you and will be less critical because they know you will listen to their concerns and address them quickly.

It may cost you a little to have someone come out and take care of a small problem, but in the long run it usually costs less, especially when you factor in the cost of an unhappy customer.

Conflict Resolution

Conflicts arise among the various parties on construction projects nearly every day. The superintendent is called upon to arbitrate many of these disputes. For example, if one trade contractor has damaged the work of another, it is up to you to make sure that a fair and timely resolution to the problem is achieved.

Disputes with buyers are somewhat different in that the superintendent has a greater vested interest in the outcome. You will find that establishing and following procedures, particularly the types of procedures discussed in this book, are the easiest ways to avoid or settle these disputes.

Handling Buyer Conflicts

When you find yourself involved in a conflict with a homeowner, keep your head and act in a professional manner. The skill of listening is essential. It is important to lis-

---

Deal Promptly with Homeowners' Concerns

One builder paid a huge price when a homeowner's concerns were put off. The homeowner raised concerns about the quality of the framing. The superintendent decided to let the problems go until the framing punch-out. The superintendent became very busy on other jobs, however, and failed to make sure the items were addressed before the insulation and drywall were installed. The homeowner became very upset and ordered the builder off of the job. Then the homeowner called the Better Business Bureau, the state contractor's licensing division, the local newspaper, and the local television station.

An investigation determined that the framing problems had not been corrected. Much of the drywall had to be removed in order to go back and fix the problems. The state threatened to pull the contractor's license and the TV station ran a segment on the evening news about unscrupulous contractors, featuring the builder's project.

The superintendent was fired, the builder spent over $13,000 correcting the problem, and the company's reputation suffered a tremendous blow. All of this over a few relatively simple framing problems that could have been corrected easily in one afternoon by one good framer. As an interesting postscript, the customer indicated to the production manager (who took over the job after the superintendent was fired) that the list the production manager made of the problems on the job was the first time anyone in the company had cared enough about the customer's concerns to write them down.

ten with an empathetic ear so as to understand the position of the homeowner. First, find out what the person's concerns are. Make sure that you understand the real problem. Sometimes people buying a home speak of symptoms and not problems. For example, one homeowner seemed to be extremely concerned over the progress of construction on his home. Even minor delays seemed to really upset him. When the superintendent got to the bottom of the issue, he discovered that the homeowner had absolutely no savings and was concerned about being out of his apartment on an exact date in order to save the last month's rent and be able to pay the closing costs on the mortgage. Once the superintendent understood the real problem the solution was relatively simple. Another buyer had just the opposite problem. He was constantly delaying the project and calling a complete halt to the work. At one point he stopped the job for over two months during the middle of the winter to "check the deflection of the foundation after it was backfilled." When the builder got to the bottom of the problem he discovered that the buyer had a number of very lucrative investments that were doing very well and that the longer he delayed payment to the builder the more money he made on his investments.

Once you understand the position of the buyer, be sure you also understand the position of the builder before you offer a solution. State your position in clear and simple terms, giving all the reasons or justifications for your position. If your specific contract language or policies have been conveyed previously to the homeowner, reiterate them in a tactful, non-confrontational way. You must then evaluate both positions in a fair-minded way. If you determine that the buyer is right, admit it and take appropriate action. However, you must remember that a fine line exists between exercising the golden rule and giving away the store. Admissions of responsibility or liability made by a superintendent in a spirit of appeasement may be used later in a courtroom to win an otherwise weak case.

If the homeowner's position is weak or based on faulty reasoning, tactfully point out the fallacies in the homeowner's argument. Take care to use thoughtful words in these instances in an effort to smooth an already sensitive situation. Cramming something down a homeowner's throat may give the superintendent some momentary emotional satisfaction, but it brings a high price in lost goodwill and damaged reputation. Through some quick but creative thinking you often can come up with a win-win solution, one where the homeowner and the builder both come out ahead.

Daily Job Log. Many superintendents keep a daily job log or diary. Notes about conflicts and agreed-upon solutions should be entered promptly and exactly, as notes made at the time of the occurrence will carry a great deal more weight in the unhappy event that you end up in court. While no one plans to end up in court, it only takes one experience there to make you realize the importance of proper documentation. Keeping a daily job log is time-consuming and sometimes inconvenient. Some superintendents prefer to carry a small tape recorder and a tape for each job. As the project progresses, the superintendent records pertinent information on the tape. A tape recording may not have the same degree of legal validity as a written record but it is not a bad substitute and is relatively painless.

Completion

Completion of the home can be complex process. Many things must be done or obtained in a short period of time, including the following:

- homeowner walk-through and maintenance training
- punch list completion
- final inspection
- health department approvals (septic tank and well)
- certificate of occupancy
- final surveys
- lending institution final inspection and approval
- final payment
- construction loan payoff and closing
- permanent loan closing
- hookup and transfer of utilities to the owner
- warranty information conveyed to the owner
- transfer of insurance coverage to the homeowner
- keys given to the owner
- owner move-in

All of these items require a considerable amount of coordination. A completion and closing checklist can be a useful tool to help you make sure all items are completed in their proper order (see Figure 7.2). All of the punch-list items should be completed prior to closing.

After Closing

After the completion of the home is another critical time in a builder's relationship with the homeowner. The homeowner's demands at this point will not always coincide with builder policy. However, effective, common-sense procedures combined with tact, good listening skills, and a willingness to please, can go a long way toward eliminating many of the disagreements common in post-sale customer service work.

FIGURE 7.2 Completion and Closing Checklist

Job number: _____   Name: _____   Date: _____

- ☐ Final cleanup
- ☐ Pre-closing inspection (in-house; home checked against purchase agreement, all change orders, inspection checklist)
- ☐ Order settlement statement, including reconciliation of all allowance monies
- ☐ Final building inspection by local authority
- ☐ Correction of deficient items
- ☐ Certificate of completion
- ☐ Certificate of occupancy
- ☐ Insurance transfer to final homeowner's insurance (certificate required)
- ☐ Water test, well certificate (where required)
- ☐ Septic system inspection, certificate (where required)
- ☐ Final survey (where applicable)

- ☐ Homeowner walk-through and orientation
- ☐ Warranties conveyed to homeowner
- ☐ Warranty service policies and procedures (review)
- ☐ Punch list complete
- ☐ Lending institution final inspection
- ☐ HUD, FHA, VA final inspection (where applicable)
- ☐ Final closing scheduled with attorney, title company, or appropriate party
- ☐ Construction loan payoff
- ☐ Lien releases (obtained as required)
- ☐ Homeowner satisfaction survey completed, returned
- ☐ Final closing with homeowner
- ☐ Final payment received
- ☐ Keys conveyed to homeowners
- ☐ Homeowner move-in

Superintendent's signature: _____   Date: _____

Customer Service Is Critical, Even after Closing

At one homeowner focus group various builders serving one area were discussed openly. One builder in particular became the topic of much of the conversation. Every one of his buyers indicated that they were really excited about their new home and were happy with the overall quality. However, they all were extremely critical of the builder's lack of response to their customer service concerns.

These homeowners indicated that once they had closed on the home and the builder had been paid, he ceased to care at all about their problems. He would not return their phone calls. He made promises to fix things and never showed up. When homeowners took time away from work for appointments with repair people they were often stood up. Repairs regularly consisted of patch work rather than thorough, proper corrections.

At the meeting these homeowners indicated that—despite their satisfaction with the overall quality of their homes—they wished they had not bought from this builder and they had already told their friends not to buy from him. The message from this focus group was clear: Even after closing, customer service is critical!

One of the keys to successful homeowner relations after completion is a written warranty. Like other documents intended for customers, the warranty need not be written in a complex legal style. And it need not increase the obligations of the builder. In many locations builders are required by law to warrant a home for one year. It is better to have a written warranty agreement signed by both parties that spells out exactly what is covered under the warranty than to have continual disagreements about what is covered and what is not.

Many builders give the client a copy of the warranty agreement when they sign the purchase agreement and review the warranty during the preconstruction conference. That way the customer knows what to expect in warranty procedures before construction ever begins. This practice eliminates the traditional, "I did not know that wasn't covered" that is common in construction service work. However, even when warranty information is reviewed up front, another review just before closing helps ensure the requirements, policies, and procedures are properly understood and fresh in the mind of the homeowners.

The trade contractors and product manufacturers may also warrant work and materials involved in roofing, siding, and systems in the home such as the furnace, water heater, and cabinets. During the final walk-through with the homeowner these warranties should be explained, including the procedures for registering the warranty (with manufacturers) and for requesting warranty service if a problem occurs.

Scheduling Service Calls. Random, spur-of-the-moment service calls waste time and money. If the construction superintendent has the responsibility of handling service calls, the superintendent should arrange to have them scheduled on a regular, manageable timetable. Instructions for scheduling service calls should be put in writing and the procedures should be made clear to all homeowners—by the sales staff, the customer service staff, and the superintendent.

Scheduled post-closing customer service is usually based on two routine calls: one at 30 to 60 days and another after one year. Buyers are asked to put together a list of minor problems for correction at these scheduled points. Many companies increase buyer goodwill by performing these service calls even when there are no obvious repairs to perform, dropping by simply to answer questions and explain any unclear operating procedures. Of course emergency service should be provided immediately on an as-needed basis.

Trade Contractors and Customer Service. If you are a superintendent in a small company, chances are you depend on your trade contractors to execute most of your service work. Excellent communication skills and effective record-keeping are the superintendent's first lines of defense in ensuring timely, quality customer service work from trade contractors. Do not shift a customer's complaint to a trade contractor just to get rid of it. Instead, direct service calls to the proper trade contractor when appropriate. Your company may wish to consider using standardized four-part forms to write up work orders for trade contractor callbacks. Keep one copy and give the rest to the trade contractor; upon work completion, one copy remains with the trade contractor, one copy remains with the customer, and one copy comes back to you (or the customer service office, if applicable).

Apart from the forms, keep a log of customer service work. When was the complaint received? Who was assigned to do the work? When was it scheduled to be completed? When was it actually completed? How long did it take from the time it was reported until it was actually complete? How much did it cost? In your weekly production meeting the status of customer service work should be a major topic of discussion. The status of each customer service item should be discussed and plans made for taking care of the work as soon as possible. Emergency work should be given top priority—even priority over new construction.

A superintendent's careful monitoring of customer service work that ensures prompt, reliable service can become a real strong selling point for the building company. Remember, the customer does not care *who* does the work—only that it gets done, properly and quickly. You may therefore wish to form an agreement with your trade contractors that asks them to assign certain regular times to perform customer service work.

Warranty Service Voucher System. Many building companies offer their customers a special voucher system that provides for callback service while providing an incentive for homeowners to perform their own small repairs rather than request service callbacks. Such a system begins with an account opened in the name of the buyer for a pre-determined amount, such as $200. Special vouchers are then given to the buyer, each worth $20, for use in requesting callback service. Service work is then paid for with the vouchers and any funds remaining in the account at the end of the warranty period are returned to the buyer, with interest. Warranty service voucher programs can be extremely successful and may prove effective in your company if carefully managed.

Carol Smith, the nation's leading expert on home builder customer service, has written an excellent two-volume *Customer Relations Handbook for Builders*, published by NAHB's Home Builder Press (see Additional Resources). This manual will be very helpful to any builder interested in establishing an excellent customer relations program. The handbook is full of checklists, forms, letters, and tools to help you develop a world-class customer service program. It is very well written and comprehensive.

8

Safety Management

Construction is inherently a dangerous business and has traditionally been a difficult industry in which to work. Work is often performed under adverse weather conditions and involves the use of hazardous materials, tools, and equipment. Other hazards include noise, dust, excavation cave-ins, and the potential threat of falling or being struck by falling material and equipment. The resulting injuries can prove catastrophic in terms of loss of life and personal injury.

Each year more than 7,000 people die in construction-related accidents. In 1994 the construction industry had the greatest number of fatalities (1,027) of any industry. That is an average of three workers killed each day. The construction industry employs more than five million workers—that's more than 5 percent of the American workforce. But more than 10 percent of industrial accidents happen in this industry and more than 19 percent of all work-related fatalities occur on construction jobs.

Over 300,000 lost-time accidents take place in the construction industry each year. In 1970 the medical cost directly associated with workers' compensation was $700,000. By 1994 the cost had risen 27 times to over $18.9 billion. Work-related accidents in construction average 54 percent higher than in other industries. The cost of accidents in our industry is more than 6.5 percent of all construction dollars expended.

As these statistics show, the cost of injuries due to accidents and neglect of safety rules is tremendous. Yet this problem is manageable. The overall cost of a safety program typically runs about 2.5 percent of direct labor costs. Interestingly, builders with accident-prevention and safety programs have 36 percent lower accident rates.

Accident Prevention Saves Money

One builder who was building about 800 homes per year in 1995 had a workers' compensation bill of over $750,000 each year (about $950 per house). When this builder decided to take safety seriously and focus on accident prevention and safe work habits, the company's accident rate dropped tremendously. The company was able to reduce its workers' compensation premiums to under $500,000 over a three-year period. During the same period of time, the number of superintendents employed by the company tripled from 31 to over 100 and the number of houses produced increased 2.7 times. The company's workers' compensation costs dropped to about $225 per house.

Safety issues affect the bottom line of every construction company, even those that subcontract all of their work to qualified trade contractors. Safety-conscious trade contractors pay less in workers' compensation premiums, have a lower rate of employee turnover, are more efficient, productive, and professional, and make higher profits. As one contractor commented, "Production is improved because the safety planning process means that you have to think through the work you are going to perform." And yet a safe trade contractor usually doesn't cost any more than an unsafe one.

Recent statistics from the U.S. government include some encouraging signs that the industry's safety record is improving. The construction industry's accident and illness rate dropped 26 percent between 1990 and 1995 and 14 percent from 1994 to 1995. Remarkably, these improvements took place during a period of economic growth when construction activity was increasing dramatically. One recent insurance company study indicated a 33 percent decrease in the frequency of lost-time claims for construction-related accidents since the early 1990s.

Most safety experts credit the dramatic decrease in accidents to increases in safety training. Safety training has become standard practice on many construction jobs and residential builders have realized the importance of safety training in the effective management of their businesses.

Safety is a company-wide concern, but as the superintendent you are directly responsible for employees' safety on the jobsite. You also are responsible for making sure that your trade contractors are properly trained and are working in a safe manner.

Three Reasons for Safety

There are humanitarian, economic, and legal reasons to practice and enforce jobsite safety.

Humanitarian Reasons

The primary reason to require safe practices on the jobsite is simply because you care about your employees, your trade contractors, and their employees. It is a traumatic experience to face informing the spouse or relative of one of your workers that their loved one has been seriously injured or killed in a jobsite accident. The spouse or relative is devastated, but so is the individual—typically the superintendent—who was responsible for safety on the jobsite.

Your most important resource is the people who work for you. If they are your most important resource, protect them! You don't leave dollar bills lying around the jobsite to blow off in the wind or get stolen. Why tolerate undue risks to employees or trade contractors when they are worth infinitely more than money? No job is so urgent that we must compromise the safety of the workers to get it accomplished.

How many times have you been required to walk on top of an exterior wall, sometimes two or more stories in the air? The danger is obvious. Yet we allow workers to expose themselves to unnecessary risks every day. Often we assume that it is OK because "accidents don't happen to us; we are more careful than other people." Yet if you look closely you will find that you are not more careful, just more lucky. The people on your jobsites are simply too important to allow them to risk injury, disability or even death.

Economic Reasons

The cost of workers' compensation has skyrocketed in recent years. Workers' compensation premiums in the construction industry typically range from a low of about 7 percent of direct wages to as much as 60 percent for some trades. The rate varies from state to state, but the rate for rough carpenters is typically 11 percent. Roofing is about 20 percent. That means that for every dollar you pay to your employee in wages you must take 11 to 20 cents out of your pocket to pay for workers' compensation insurance. It does not matter if the employee is on your payroll or is a trade contractor's employee. The same amount must be paid. On an average 2,000-square-foot house, the cost of workers' compensation insurance covering your workers and the employees of your trade contractors is $4,321. That exceeds the profit—and often the rest of the overhead—for the entire job.

Workers' compensation comprises two main parts, the *manual rate* and the *experience modification rate* (EMR). The manual rate is the rate per $100 of payroll paid by the contractor for workers' compensation for each trade. The rate varies with the potential hazards and historical claims associated with each particular job classification. The manual rate for carpenters is often about $13 to $15 per $100 of payroll or between 13 percent and 15 percent. Roofers are typically in the $20 range.

Every company in the U.S. that has employees has an EMR. The EMR is a result of a complex formula that takes into account all of the accidents and workers' compensation claims of each company. In essence, the rate is based on a comparison of the actual losses for a particular company against the expected losses. The EMR is based on the losses for a three-year period. The most recent year is excluded in the calculation because often only partial information is available for the most recent year. So in the year 2000 your EMR will be based on your loss ratio for 1998, 1997, and 1996. If your actual losses for this three-year period are about the same as your expected losses, you will have an EMR of 1.00. If your actual losses are higher than your expected losses, your EMR will be greater than 1.00—for example, 1.70. If your actual losses were less than your expected losses, your EMR will be less than 1.00—for example, 0.65.

The EMR also takes to consideration the frequency of accidents. A company that has one accident resulting in a single large claim will have a lower EMR than a company that has many accidents resulting in small claims. The reasoning behind this is that a large number of accidents reflects an unsafe company, and any of the accidents could have resulted in a large claim.

Your workers' compensation premium is based on the following formula:

$ Payroll x Manual Rate x EMR = Workers' Compensation Premium

Any building company's public image can be seriously damaged or even destroyed by a single accident resulting from careless and unsafe conditions on a jobsite. It can prove almost impossible for a builder to live down the poor reputation resulting from a needless fatality due to an unsafe condition. In addition, the adverse publicity can be personally devastating. For example, at one construction site, a five-year-old boy drowned in a water-filled hole dug for a septic tank. The resulting lawsuit and negative publicity were factors in the construction company's later declaration of bankruptcy. When a fatality is involved this result is not uncommon. Moreover, the personal grief suffered by persons who are witness to or involved in a fatal accident never truly ceases.

A Good Safety Record Makes a Big Difference

An example may be helpful in illustrating the impact a jobsite safety program can have on the cost of workers' compensation premiums. If one company has a good accident claims record resulting in an EMR of 0.65 and another company has a poor record resulting in an EMR of 1.7, how will it affect their workers' compensation costs?

	Company A	Company B
Payroll	$100,000	$100,000
Manual rate	x 0.15	x 0.15
	$ 15,000	$ 15,000
EMR	x 1.70	x 0.65
Workers' compensation Premium	$ 25,500	$ 9,750

As you can see, a good safety record can make a big difference.

Legal Reasons

The legal requirements for pursuing jobsite safety have two aspects: the potential for liability and compliance with requirements of the Occupational Safety and Health Administration (OSHA).

Liability Issues. The builder's potential for liability in the event of an accident is tremendous. Any injured worker is a potential financial time-bomb. High monetary awards given to workers, jobsite visitors, and others who are injured, disabled, or killed on construction jobsites have caused builders to become much more aware of potential jobsite risks. Increasing medical and insurance costs, coupled with the high compensatory and punitive damages from sympathetic juries, have resulted in tremendous potential liability. As a result liability and workers' compensation insurance premiums have risen on average 15 percent per year for the last 20 years—fully three to four times the rate of inflation.

OSHA. In 1970 Congress passed the Occupational Safety and Health Act (Occupational Safety and Health Act). Nobody could have foreseen the affect this legislation would have on the construction industry. In a recent survey by *Builder* magazine, builders were asked to name their greatest concerns. Development costs, increased lumber prices, and a shortage of trained workers were high on the list. But the number one concern builders mentioned was workers' compensation, and the number two concern was OSHA inspections. Another survey conducted by *Nation's Building News* showed the number one concern was OSHA regulations, while safety and health legislation ranked twelfth.

A serious violation of an OSHA regulation can result in fine of as much as $7,000 per violation. A repeat violation of an OSHA regulation can result in a fine of as much $70,000 per violation. Residential builders have received fines in excess of $300,000 on a single project. That is enough to get the attention of any builder!

Larger builders have become very aggressive in implementing safety programs and small-volume builders are beginning to acknowledge the importance of jobsite safety to the successful operation of their businesses.

Before 1970 and OSHA, the enactment of safety and health laws had been left to the individual states. OSHA was specifically charged to accomplish the following things:

- promulgate, modify, and revoke safety and health standards
- conduct inspections and investigations and issue citations, including penalties
- require that records be kept

- restrain imminent danger situations (through the courts)
- approve or reject state safety plans
- provide education and training to employers and employees
- consult on prevention of injuries and illnesses
- grant funds to states for the development and operation of state safety plans and programs
- develop and maintain an occupational safety and health statistics program

Ten regional offices have the primary mission of supervising, evaluating, and executing all programs of OSHA in their regions. Each region has *area offices,* headed by area directors. Area offices carry out the compliance aspect of the OSHA programs within their geographic areas.

The Occupational Safety and Health Act encourages the states to assume responsibility for the administration and enforcement of their own occupational safety and health laws. The following states and possessions have approved plans as of 1998:

- Alaska
- Arizona
- California
- Hawaii
- Indiana
- Iowa
- Kentucky
- Maryland

- Michigan
- Minnesota
- Nevada
- New Mexico
- North Carolina
- Oregon
- Puerto Rico
- South Carolina

- Tennessee
- Utah
- Vermont
- Virgin Islands
- Washington (state)
- Wyoming

To be approved, a state plan must be "at least as effective in providing safe and healthful employment" as the federal program.

Under the Occupational Safety and Health Act each employer covered by the act has the general duty to "furnish places of employment that are free from recognized hazards that are causing or likely to cause death or serious physical harm." This language is commonly called the *general duty cause.* Congress has authorized, and OSHA now provides through a state agency or private contractors, free on-site consultation services for employers in every state.

Employees cannot be required to perform work under conditions that are unsanitary, hazardous, or dangerous to their safety or health. Each employee has a duty to comply with the safety and health standards and all rules, regulations, and orders that are applicable to his or her own actions and conduct on the job. Employees also have the right to information from the employer regarding the toxic effects, conditions of exposure, and precautions for safe use of all hazardous materials in the establishment. Employees must be given access to records of their history of exposure to certain toxic materials or harmful physical agents. Employees have the right to review the Log and Summary of Occupational Injuries (OSHA FORM 200) at reasonable times and in a reasonable manner. Employees have the right to confer in private with the Compliance Safety and Health Officer (CSHO, or simply compliance officer) and to respond to questions from the compliance officer in connection with an inspection of an establishment. An employee also can place an anonymous written request to OSHA to make a special inspection if the employee believes a violation of a standard threatens physical harm. Employees must be given the opportunity to review citations given to an employer, and the citations must be posted at or near the place where the violation occurred. Employers are required to prominently display the OSHA poster (OSHA 2203) in a conspicuous place in the workplace where notices to employees are customarily posted. The poster informs employees of their rights and responsibilities under the act.

Employers also are required to "instruct each employee in the recognition and avoidance of unsafe conditions and the regulations applicable to his work environment to control or eliminate any hazards or other exposure to illness of injury." Employers must maintain records of safety training and of occupationally related illnesses and injuries. They must notify the OSHA area office either orally or in writing within 8 hours after the occurrence of an accident that is fatal to one or more employees or that results in the hospitalization of three or more employees.

OSHA Inspections

OSHA has been given the authority to make inspections to ensure compliance with the Occupational Safety and Health Act. An employer representative as well as an authorized employee representative may accompany the compliance officer during the official inspection of the premises and all its facilities, and may participate in the closing conference. This inspection should include an opening conference that will indicate the reason for, nature, and extent of the inspection. The closing conference is designed to give the builder preliminary results of the inspection. No citations will be issued and no fines will be levied at the closing conference; however, the compliance officer will advise the employer and the employee representative of any conditions and practices that may constitute a safety or health violation. The compliance officer also should indicate the applicable section or sections of the standards that may have been violated. The compliance officer normally will advise that citations may be issued for alleged violations and that penalties may be proposed for each violation.

Administratively, the authority for issuing citations and proposed penalties rests with the area director or his representative. The compliance officer should also inform the employer that any citations will establish a reasonable time for abatement of the alleged violations and explain the appeal procedures with respect to any citations or notices of a proposed penalty. Employers and employee representatives have the right to participate in both the opening and closing conferences.

Inspection Priorities

OSHA has established priorities for the assignment of agency personnel and resources. The priorities currently are as follows:

- investigations of imminent dangers
- catastrophic or fatal accidents
- investigations of employee complaints
- programmed high-hazard inspections
- re-inspections and follow-up inspections

Investigations of Imminent Dangers. The Occupational Safety and Health Act defines *imminent danger* as "any condition or practice in any place of employment which is such that a danger exists which could reasonably be expected to cause death or serious physical harm immediately or before the imminence of such danger can be eliminated through the enforcement procedures otherwise provided by this Act." *Serious physical harm* includes the permanent loss or reduction in efficiency of a part of the body or inhibition of a part of the body's internal system such that life is shortened or physical or mental efficiency is reduced. Allegations of an imminent danger situation will ordinarily trigger an inspection within 24 hours of written notification. In addition, an OSHA officer is likely to stop if he drives by a jobsite and sees a situation that appears to constitute an imminent danger.

Catastrophic Accident or Fatality. Employers are required to notify OSHA within 8 hours of a fatality or catastrophe. A *catastrophe* is defined as the hospitalization of three or more employees.

Investigations of Employee Complaints. Highest priority is given to written complaints that allege an imminent danger situation.

Programmed High-Hazard Inspections. OSHA selects industries or activities for inspection based on the death, injury, and illness incidence rates, employee exposure to toxic substances, and other criteria. In residential construction OSHA has targeted inspections to focus on trenching and fall protection, two high-hazard areas.

Re-inspections. Establishments cited for alleged serious violations normally are reinspected at some point to determine whether the hazards have been abated.

Followup Inspections. Followup inspections always will be made for situations involving imminent danger or where citations have been issued for serious, repeated, or willful violations.

Violations, Citations, Penalties, and the Appeal Process

The occupational safety and health standards promulgated under the Occupational Safety and Health Act also can be used as a basis for determining alleged violations. When an investigation or inspection reveals a condition alleged to be in violation of the standards (or the general-duty clause), the employer may be issued a written citation that describes the specific nature of the alleged violation, the standard allegedly violated, and that fixes a time for abatement. The employer must post the notice of a citation in a "prominent place" until the violation has been corrected or for a period of three days, whichever is later. Failure to do so can bring a penalty of up to $7,000. Again, all citations are issued by the area director or his representative. Citations are sent to the employer by certified mail.

Types of Violations and Severity of Penalties

OSHA has defined four types of violations and two additional penalties. For some kinds of violations the area director has discretion to adjust the dollar amount of penalties based on factors such as an employer's good faith (efforts to comply with the act), history of previous violations, and size of business. The severity of penalties is based on several factors, including the likelihood that the violation will result in injury or illness, the likely severity of the injury or illness (should one occur), and the extent and frequency to which the OSHA standard has been violated.

Other-than-serious Violations. Violations deemed "other than serious" have a direct relationship to job safety and health but probably would not cause death or serious physical harm. The proposed penalty for such penalties (up to $7,000 per violation) is discretionary. The area director may adjust the penalty downward by as much as 95 percent, depending on the employer's good faith, history of previous violations, and size of business.

Serious Violations. A serious violation is a situation involving substantial risk of death or serious physical harm and about which the employer knew, or should have known, of the hazard. The mandatory penalty (up to $7,000 per violation) may be adjusted downward based on the employer's good faith, history of previous violations, the gravity of the alleged violation, and size of business.

Willful Violations. A violation is categorized as willful if the employer knowingly commits the violation or commits the violation with plain indifference to the safety and health of employees. In other words, the employer either knows he or she is in violation of an OSHA standard or is aware that a hazard exists but makes no reasonable effort to eliminate it. A minimum penalty of $5,000 for each violation and penalties may go as high as $70,000 per violation. A proposed penalty may adjusted downward (but not below the minimum), depending on the size of the business and its history of previous violations. Usually no credit is given for good faith. If an employer is convicted of a willful violation that resulted in the death of an employee, the offense is punishable by a court-imposed fine of up to $250,000 for an individual or $500,000 for a corporation and up to six months' imprisonment for a criminal conviction.

***De Minimis* Violations.** *De minimis* violations have no immediate or direct relationship to safety or health. *De minimis* is short for the legal maxim, *De minimis non curate lex,* "The law does not concern itself with trifles." Although *de minimis* violations will not be included in the citation, if they are found during an inspection they will documented the same as other violations. If a builder is later found to have other violations, the record of *de minimis* violations may make it more difficult for the builder to demonstrate a history of good faith.

Repeat Violations. A violation is considered a repeat violation if an earlier citation was issued for a substantially similar violation of a given standard or for the same condition that violates the general-duty clause. A repeat violation applies to the company, not to the job or jobsite. This means the company may be cited for a repeat violation if a substantially similar violation has been found to be present on different sites (even in different states). A repeat violation can bring a penalty of up to $70,000 per violation.

Failure to Abate a Prior Violation. When cited for a violation the employer must correct the hazard by the prescribed date or petition for an extension. Failure to abate a prior violation within the prescribed time period can result in a civil penalty of up to $7,000 per day for each violation.

Falsifying Records, Reports or Applications. Falsifying information can result in a penalty of up to $10,000 and up to six months' imprisonment.

OSHA Focused Inspection Program

In the United States 90 percent of all construction workplace fatalities are caused by four types of accidents and injuries:

- falls (33 percent)
- being struck by materials, equipment, or other objects (22 percent)
- being crushed or trapped, as in excavation cave-ins (18 percent)
- electrical shocks (17 percent)

In 1994 OSHA implemented a focused inspection program. In accordance with the program compliance officers now perform comprehensive inspections only at jobsites where the contractors exhibit little or no apparent commitment to safety. Contractors who show a demonstrated commitment to safety receive a focused inspection that concentrates on the four primary hazards mentioned above. To qualify for a focused inspection the contractor must meet the following criteria:

■ The controlling contractor (the builder) must have an effective safety program. The compliance officer will ask to see a copy of the builder's safety program and review it carefully. The compliance officer also will interview employees to determine whether the safety program is actually used and effective. Simply having a program is not enough. OSHA will look for "vitalized, implemented programs," not just dusty notebooks sitting on a bookshelf.

■ The builder must have a competent person responsible for managing safety on the jobsite. *Competent* means this person must be able to identify existing and predictable hazards in the surroundings, or working conditions that are unsanitary, hazardous, or dangerous. This competent person must have the authority to take prompt corrective action as needed to eliminate them.

If the builder meets both of these criteria the compliance officer will conduct a focused inspection of the jobsite, concentrating on the four primary groups of hazards. If the inspection results from an employee complaint or from an observed imminent danger, the inspection will also include the items of concern. If in the process of the inspection the compliance officer notes any other serious hazards in plain view that pose an high probability of death or serious physical harm, these hazards may also be cited—or the inspection may revert to a comprehensive inspection. However, if the hazards are abated immediately and the compliance officer observes the abatement, they may not be cited. Other-than-serious violations normally will not be cited as a result of a focused inspection unless there are so many other-than-serious violations the compliance officer concludes that the safety program is ineffective. The focused inspection program in effect rewards builders who make a good-faith effort to develop effective safety programs.

What to Do in an OSHA Inspection

When you find yourself involved in an OSHA inspection, take the following steps:

1. Notify your office when the OSHA compliance officer (compliance officer) arrives.
2. Determine if a warrant should be required. Legally, OSHA may not conduct an inspection without a warrant without the employer's consent. Keep in mind that failure to admit an inspector may arouse suspicions and create an unfavorable impression.
3. Ask to see the inspector's credentials. Record the inspector's name, serial number, and the name of his or her supervisor.
4. During the opening conference, learn the purpose and scope of the inspection. If an employee complaint or drive-by observation by an inspector triggered the inspection, ask for copies of the applicable safety and health standards and for a copy of the complaint.
5. Be sure your OSHA poster is up, your assured grounding material is available, your hazard communication program is on hand, and your lockout procedure is available, if applicable.

6. Do not disclose any information that can be used against you. Do not ask the inspector if something is or is not in compliance. Be polite, yet firm. Do not hesitate to boast about the safety precautions taken on the project.

7. During the walk-around inspection make sure a supervisory employee accompanies the inspector. The inspector has the right to consult with a reasonable number of employees concerning safety and health matters. The inspector can talk with your employees in private and ask for their home telephone numbers. However, workers are not obligated to discuss anything with OSHA.

8. Take photographs of the same items photographed by the inspector.

9. Make a note of every violation the inspector points out. If possible, correct all violations on the spot and be sure the inspector's records reflect the correction.

10. Note the amount of time the inspector spends on the jobsite.

11. Trade contractors should be present during inspections of their work.

12. During the closing conference, go over every item for which you will be cited. Check the item against the standard. Ask for a complete explanation if any item is not absolutely clear. Should you disagree with the inspector's position, politely yet firmly point out your opinion.

13. Do not discuss the fine! This could be construed as a bribe.

14. Expect a follow-up inspection if your citation is classified as serious, repeat, failure-to-abate, willful, or if the safety violation is egregious.

Appeals

Following issuance of an OSHA citation the employer has the right to contest the citation. Generally speaking it is wise to automatically contest all OSHA citations. The first step in contesting an OSHA citation is to request an informal postinspection conference. Held by the OSHA assistant regional director, the informal post-inspection conference provides a venue for discussion of issues raised by inspections, citations, proposed penalties, or a builder's notice of intent to contest.

The conference may be requested by either the employer or the employee representative, and both parties shall be afforded the opportunity to participate fully. The employer presents his or her case. Input often is sought from employees if represented at the conference. Considerable negotiation often takes place, the intent being the overall improvement of safety on the jobsite. Fines normally are reduced of the employer can demonstrate a good faith effort, and in fact most OSHA citations are resolved at the informal hearing.

An employer also can file a notice of contest with the OSHA area director. The notice must be filed within 15 days of the citation. There is no specific format for the notice, but a copy of the notice of contest must be given to employees. The notice of contest is then forwarded to the OSHA Review Commission (OSHRC), an independent agency comprising three commissioners who are not affiliated with the Department of Labor or with OSHA.

The case may be assigned to an administrative law judge, in which case a hearing usually is scheduled near the employer's workplace. Employers and employees both have the right to participate in the hearing. Representation by an attorney is not required.

After hearing the case, the administrative judge rules on the case. Further appeal may take place before the OSHRC. Commission decisions may be appealed to the U.S. Court of Appeals.

Affirmative Defenses

The more common defenses for an OSHA citation include the following:

- The employer did not create the hazard.
- The employer did not have the responsibility or the authority to have the hazard corrected.
- The employer did not have the ability to correct or remove the hazard.
- The employer can demonstrate that the subcontractors (trade contractors) who had a role in creating a hazard or exposing their employees to a hazard have been specifically notified of the hazard, and that any employer (including any trade contractor) who has a role in correcting the hazard has also been notified of the hazard.
- The employer has instructed his or her employees to recognize the hazard and has informed them how to avoid the danger associated with it.
- The employer has taken appropriate alternative means to protect employees from the hazard or has removed his or her employees from the jobsite to avoid the hazard.
- The violation resulted from unpreventable employee misconduct or from an isolated event. If misconduct was involved, the misconduct was unknown to the employer and in direct violation of an adequate work rule which had been effectively communicated and enforced.
- Compliance with the requirements of the standard being violated is functionally impossible or would prevent performance of required work and no alternative means of employee protection exist.
- Compliance with the standard being violated would result in greater hazards to employees than noncompliance, no alternative means of employee protection exist, and application for a variance to the standard would be inappropriate.

MultiEmployer Worksites

It is OSHA's policy that "it is not sufficient to hold responsible only those employers who have employees exposed to unsafe conditions or unsafe practices." "Employers who have contractual responsibility for site safety and those who cause unsafe conditions are also responsible." On multi-employer work sites, citations may be issued to any of the following parties:

- the *controlling contractor* (the general contractor or builder in charge of the project)
- the *creating contractor* (the contractor who caused the hazard or violation)
- the *correcting contractor* (the contractor who had the responsibility, authority, and ability to correct the problem)
- the *exposing contractor* (the contractor who exposed his or her employees to the hazard)

The Superintendent's Role in Safety

This chapter presents a great deal of information about OSHA inspections and compliance because the superintendent must be prepared to handle the situation when an inspection occurs. However, it is important to remember that OSHA compliance alone is *not* a safety program. Mere compliance with the requirements of the act will not achieve optimum employee safety and health.

Important elements of a complete safety program, such as establishment of safe work procedures to limit risk, supervisory and employee training, and job safety analysis have a significant and far-reaching impact on overall job safety. A violation of a standard is only symptomatic of something wrong with the safety management system. Only a complete occupational safety and health program can achieve a level of risk that is acceptable to employers as well as employees. The real objective and the purpose of the Occupational Safety and Health Act—better occupational safety and health performance—extends beyond mere compliance with yet another promulgated set of standards.

Safety and health are as much a part of project planning and control as any other aspect of construction. A superintendent has primary responsibility for ensuring that construction workers have a safe and healthy place in which to work. Through OSHA this responsibility has now been mandated by law and is required by government regulations. The intent of Congress was to establish uniform standards throughout all industries to ensure and mandate adequate safety on the job. According to the act, employers "shall furnish a place of employment which is free from recognized hazards that are causing or are likely to cause death or serious physical harm to his employees." As a manager, the superintendent has a key role in making sure the employer (the builder) fulfills this requirement.

The Code of Federal Requirements (CFR) 1926, Occupational Safety and Health Standards for the Construction Industry, also indicates that, under the act, "... it shall be a condition of each contract which is entered into... that no contractor or trade contractor for any part of the contract shall require any laborer or mechanic employed in the performance of the contract to work in surroundings which are unsanitary, hazardous, or dangerous to his health or safety."

The contractor or another competent person is to conduct "frequent and regular inspections of the jobsites, materials, and equipment." The employer should also "instruct each employee in the recognition and avoidance of unsafe conditions and the regulations applicable to his work environment to control or eliminate any hazards or other exposure to illness or injury."

The CFR also states that the "prime contractor" (builder) will not be relieved of "overall safety responsibilities for compliance" with the requirements in relation to all work to be performed under the contract.

Jobsite safety begins with the individuals working on the site. Workers should begin each project believing that safety on the job is their personal concern as well as the concern of every other worker on the project. Each worker should be aware of safety hazards and be prepared to act promptly to avoid or correct jobsite hazards regardless of the existence of formal safety rules and regulations.

Like most people, however, workers often fail in this most basic form of safety management. Builders and superintendents must therefore work together to establish their own safety programs and highlight the safety requirements of the company. Figure 8.1 lists the elements of a safety program for home builders. The list is quite thorough, but you will have to adapt it to your company's specific needs. For example, if you are a developer you may need additional coverage in trench excavation.

NAHB has developed "The Model Safety Program for the Building Industry" to help you start developing a safety program for your company. The program is available on computer disk so you can easily modify the material to suit your needs. The disk comes with a manual that provides a more in depth discussion of safety management in construction. The model program also contains several forms that you may find helpful, including an accident investigation report form and safety inspection checklists.

FIGURE 8.1 Contents of a Safety Program

Table of Contents

Policy Statement on Safety

Program Safety Goals (Specific Goals and Objectives)

Organizational Safety Responsibilities
- Contractor
- Superintendent
- Trade contractors
- Employees

Safety Education and Training
- New employee orientation
- Training requirements in safety and OSHA standards
- Continual training (tailgate safety meetings)
- Training record keeping

OSHA Inspections
- Training for OSHA inspections
- Policies for OSHA inspections

Hazard Communication
- Toxic and hazardous substances
- Material safety data sheet (MSDS)
- HazCom training

General Safety Rules
- Tools and equipment
- Personal protective equipment
- Material handling and storage
- Fire prevention
- Housekeeping
- Demolition (especially for remodelers)
- Excavation hazards
- Concrete and masonry hazards
- Motor vehicles, cranes and mechanized equipment (where applicable)
- Electrical hazards
- Fall protection
- Scaffolds
- Stairways and ladders
- Other rules as needed (in welding, power transmission, and so forth)

Trade Contractors (Compliance, Training, Safety Programs, Discipline, Other Topics)

Emergency Action Plan

Lockout, Tag Out

Internal Safety Inspections

Accident Investigation
- What to document
- Supervisor training

First Aid
- Training
- Procedures

Substance Abuse

Record Keeping and Documentation
- Purposes
- Employer's first report of injury or illness (OSHA number 101)
- Log and annual summary of occupational injuries and illnesses (OSHA number 200)

Confined Space Entry Policies and Training (Where Applicable)

Enforcement, Discipline, Reward Program

Just creating a safety program is not enough. You have to really use it. Implementation is extremely important. An effective safety program requires management commitment, employee participation, jobsite hazard analysis, hazard prevention and control, and ongoing safety training. Each builder and trade contractor must be responsible for ensuring that its own workers are properly trained and not exposed to safety hazards.

If a compliance officer inspects one of your jobs the first thing he or she will do (after initial introductions) is ask to see your written safety program. The compliance officer will ask your employees (not just your superintendent) and the employees of your trade contractors about your safety program. If you do not have a copy on the jobsite, or if you don't know where it is, you will start the inspection in a very bad position. Your safety program must be an integral part of your commitment to safety; everyone on the jobsite should know what it is, where it is, and essentially what it contains.

Training for a Safe Jobsite

One of the critical elements of any safety program is training of employees and trade contractors. When the compliance officer visits your jobsite the second thing he or she will do is ask about your training program. The compliance officer will ask your employees and the employees of your trade contractors what training they have received. You will be asked to provide records indicating the dates of training sessions, the topics and who attended the sessions. Several types of training are required.

New Employee Safety Orientation

Whenever a new employee is hired you are required to provide safety training related to the type of work they will be doing. It is the responsibility of management to instruct each employee to do the following:

- recognize, avoid, and prevent unsafe and hazardous conditions connected with particular job assignments
- be aware of and understand the safety regulations applicable to particular work assignments
- document incidents

As part of the orientation you provide to any new employee before he or she starts work at any construction facility or jobsite, give the new hire information about the company's operation and policies on subjects such as substance abuse, hazard communication, safety, and health (Figure 8.2). At a minimum, the superintendent should discuss the procedures and policies, confirm that the employee understands them, and make sure the employee knows where the safety program manual is kept on site.

Every contractor performs the safety orientation differently. Most small companies must orient each individual employee when they start construction. Larger companies hold a safety orientation every week for new employees. Mid-sized companies may schedule the safety orientation monthly or periodically as needed. The safety orientation may be provided by the superintendent or a first-line supervisor, or in large companies by a safety professional. Whatever the format, new hires must come away with the feeling that safety is the first priority for all employees from top management to the worker. The new hire must leave the orientation with the following expectations about safety:

- safety is planned into every job so there will be *no injuries*
- safety rules shall be observed and enforced with *no exceptions*
- safety is the responsibility of everyone on the jobsite

A record indicating the name of the new employee, name of the person giving the orientation, the items covered during the orientation, and the date of the orientation should be kept. The completed form or checklist should be signed and dated by the employee and the supervisor, and placed in the employee's personnel file (see Figure 8.2). The superintendent or supervisor should add items to the checklist as needed to meet the specific needs of the employee, cover hazards unique to the jobsite, and discuss tasks the employee will be performing. Films, slide presentations, videotapes, or other audio-visual aids can be useful parts of the safety orientation. NAHB has developed a series of safety videotapes to assist in training new employees. The tapes are available from NAHB's Home Builder Bookstore (see Additional Resources).

FIGURE 8.2 Safety Program Orientation Checklist

Check off each item after discussing it with the new employee and providing copies of any written policies and procedures (as applicable).

- [ ] Prohibited items and conduct, such as: fighting, harassment, alcohol, guns, drugs, and so forth
- [ ] Detailed emergency and fire procedures and evacuation routes, hazard recognition and avoidance
- [ ] Detailed explanation of all site safety rules and procedures
- [ ] Disciplinary action for failure to comply with job safety and health requirements
- [ ] General dress code
- [ ] Hard hats
- [ ] Eye and foot protection
- [ ] Hazard communication program, Material Safety Data Sheets and their location
- [ ] Personal hygiene
- [ ] Personal lifting limits
- [ ] Project communication system
- [ ] Fire extinguishers
- [ ] Drinking water

- [ ] Safety equipment
- [ ] Project housekeeping, trash handling
- [ ] Respirators (where applicable)
- [ ] Care of equipment: ladders, hand tools, electric power tools, GFCIs, electric cords, air, gas and water hoses, and so forth
- [ ] Fall protection, safety harnesses
- [ ] Scaffolds: erection, use, and care
- [ ] Materials storage and handling
- [ ] Cranes and hoists
- [ ] Caution and danger signs
- [ ] Tool and equipment lockout, tag-out
- [ ] Confined space entry (where applicable)
- [ ] Vehicle safety, speed limits, personal transportation precautions, pickup truck safety
- [ ] Reporting accidents and injuries, including near misses
- [ ] Handling injuries, first aid, nurse care, doctor care, and availability
- [ ] Bloodborne pathogens

Superintendent's signature: _____ Date: _____

New employee's signature: _____ Date: _____

Current employees who transfer to new positions within the company or new employees hired form other builders with unfamiliar safety procedures, hazards, or safety concerns should complete a modified safety orientation. New workplace job safety requirements and potential hazards the worker may encounter in the new position should be explained and any necessary additional training given. You may wish to consider developing a series of short quizzes and follow-up tests to be used as part of the company's orientation program. Quizzes and tests can help you determine the effectiveness of your safety training. Any quiz or test question answered incorrectly by the employee should be noted on the orientation record form and the superintendent or trainer should make a point to review the correct answer with the employee to ensure that he or she fully understands it. If a pattern of incorrect answers appears, the superintendent or trainer will know that the orientation needs to be made more clear or more thorough in regard to those topics.

The superintendent should also follow up with the new or transferred worker to verify that safety orientation items were clearly understood. These follow-up checks help ensure the worker is performing his or her assigned tasks in a safe manner. They also help document the completion and effectiveness of your training in case of an inspection. Some companies issue a safety training card to each employee indicating that the person has completed the safety orientation.

Ongoing Safety Training

Employees should receive training on specific hazards associated with each job. For example, if the current project will have unique fall protection or excavation hazards these topics should receive specific attention at an appropriate time before or during the project. Safe work practices should be developed, implemented, and enforced for routine operations and tasks.

Regularly scheduled safety meetings and training sessions are important to the success of your safety program. Regular weekly or biweekly meetings serve as a constant reminder to employees that their safety is of the utmost importance to the company. The superintendent is the ideal person to conduct these short safety training meetings (also called tailgate meetings or tool-box talks), which are the backbone of any company's safety training program. Management support for safety can be demonstrated by taking part in safety meetings within their area of control. Training meetings also can be held in conjunction with your weekly production meeting. (If you combine meetings take care that the agenda does not run too long lest you be tempted to skip topics.) Mandatory attendance, minutes recording the attendees and the topic discussed, and short quizzes following the training all contribute to an effective, professional training experience.

Company safety rules and regulations, safe work procedures, analysis of accidents, and discussion of potential hazards all are potential topics for weekly training sessions (see Figure 8.3). Weekly safety training meetings provide you with an opportunity to point out any hazardous or unhealthy conditions or unsafe work practices you may

FIGURE 8.3 Topics for Weekly Safety Training Meeting

The following topics should be discussed and reinforced through weekly safety training meetings:

- Subjects suggested by employees
- Specific hazards present on the jobsites or unusual exposure to hazards
- Items of concern from safety inspections
- Training in safe use of specialized equipment
- Training in use of personal protective equipment (as necessary)
- Accident and injury reporting
- Analysis of any accidents, injuries, near-misses
- Safety coordination with trade contractors or builders who may be working in and around company operation
- Company safety policies
- Company substance abuse policies
- Rights and responsibilities of employees
- Safety as a part of employee performance (including incentives such as safety awards and disciplinary policies)
- Housekeeping

- Fire safety
- Any health hazards
- Hazard communication program
- Hazard recognition and avoidance
- Demolition
- Lockout, tag-out procedures
- Confined space entry procedures (where applicable)
- Electrical safety
- Fall protection
- Scaffold and ladder safety
- Cranes, rigging, and signaling
- Excavation and trenching
- Training in safe use of tools and equipment
- Heavy equipment safety
- Motor vehicle safety
- Reporting unsafe acts and conditions
- Reward and discipline program review
- Other safety concerns

have noticed. Be sure to welcome suggestions for improving the safety program from all employees and trade contractors who attend the meetings.

Some builders use color slide presentations featuring work areas in past or current jobsites. These slides are ideal for illustrating how the company is doing and identifying areas of concern that need work. Because the slides depict actual scenes from familiar jobs they also help foster a sense of involvement and immediacy more effectively than mere lectures or "canned" audiovisual aids. On occasion you may want to invite guest speakers to address your employees. Your workers' compensation carrier or an OSHA consultant can add a new perspective to safety discussions.

Tailgate safety meetings should be short (a maximum of fifteen minutes), limited to one or two topics, and *well documented*. Record the date, employees in attendance, the topics discussed, and any follow-up measures taken. For more involved training sessions additional notes are useful, including notations of any training certificates issued to employees who successfully completed the training. The superintendent should sign and date each record, and the records should be collected and retained in a safety training file for a minimum of five years.

As appropriate, copies of safety training records can be retained at the jobsite. Keeping copies onsite can be particularly useful on large projects in case of an OSHA inspection. Copies of safety training records should be available at the jobsite and provided upon the participant's request or as mandated by law.

Safety Awards Program

Some building companies have implemented safety awards programs for their employees. These awards programs often are organized by project and can be based on the number of days without a lost-time accident or based on a specific score (using a point system) for the project's safety inspection. Awards range from hats, jackets, and other clothing items to weekend trips for the worker and spouse.

Safety Inspections

The adage, "inspect what you expect" holds especially true for safety concerns. Superintendents should conduct regular, active safety inspections of construction jobsites. In addition, upper management should perform periodic safety inspections to demonstrate their commitment toward safety, assure themselves that the company is not exposing itself to undue risk, and keep abreast of the status of safety on the company's jobsites.

Figure 8.4 presents a jobsite safety inspection checklist geared to residential construction. This sample checklist is not all-inclusive but it will give you a good idea of what to look for and can provide a starting point for developing your own checklist.

FIGURE 8.4 Jobsite Safety Inspection Checklist

Adapt the items on this checklist to suit your company's policies and procedures. This checklist includes some duplication of items that are of interest in more than one category (for example, the same item may appear in both the General and Record Keeping and Documentation categories.)

General

☐ OSHA poster in visible location, accessible to all employees

☐ Minutes of jobsite safety meetings recorded and kept at the office

☐ Safety inspection reports by trade contractors prepared and kept at the jobsite

Sanitation

☐ Working toilet provided at the jobsite

☐ Adequate supply of potable water available at the jobsite

Record Keeping and Documentation

☐ OSHA poster in visible location, accessible to all employees

☐ Emergency telephone numbers located in accessible, easily located place

☐ Written safety program in place

☐ Copy of written safety program kept at the jobsite

☐ Safety program adequate overall, provides the necessary information

☐ All accidents and injuries appropriately recorded on OSHA 200 and OSHA 101 forms

☐ Safety records accessible to employees and retained for the required time as per OSHA

Hand and Power Tools

☐ All hand and power tools in good working order

☐ Hand-held power tools equipped only with constant-pressure switches

☐ Safety devices provided on air-powered tools that prevent accidental disconnection from supply hoses

☐ Air-driven nailers operating at more than 100 psi provided with safety devices on muzzle to prevent accidental discharge

☐ Tools stored in a dry, secured location on the jobsite

☐ Electrical cords (including those on tools) free of cuts and not frayed

☐ Saws guarded by equipment-appropriate guards

☐ Tools used only for their intended use

☐ Handles of hammers and other hand tools in good condition, free of cracks and splinters

Housekeeping

☐ Work areas clean and free of dangerous waste and material

☐ Trash and scrap materials removed or stacked in orderly fashion

☐ Trash and combustible material placed in containers provided for that purpose

☐ Scrap lumber, hoses, cable wiring, and all other debris cleared away from work areas, hallways, and stairways

☐ Nails removed from scrap lumber and other unused materials

☐ No spills of liquids or other materials that may cause an accident

☐ Work areas have appropriate levels of lighting (as needed)

☐ Holes and openings protected and appropriately marked

Fire Prevention

☐ Fire extinguisher (2A rating minimum) provided for every 3,000 square feet of space

☐ Portable fire extinguisher within 100 feet of all working areas

☐ Portable heaters used only in accordance with specifications

☐ All employees or trade contractors know fire extinguisher location(s) and how to operate extinguisher(s)

Personal Protective Equipment

☐ Hard hats worn in construction areas where there is a risk of injury

☐ Hearing protection worn in construction areas during periods of moderate, extreme, or long-term noise

☐ Mandatory eye protection worn in construction areas when the following tools or materials are in use: hammers (all types); saws; chipping tools; brooms; grinders; impact tools; drills; chemicals; hazardous substances that create dust, mist, or fumes; concrete (during placement); and grout

☐ Hard hats and eye protection worn by employees, visitors, and vendors in the immediate construction area

☐ Face shields worn when danger of harmful chemical or physical contact with the face is present

☐ All eye and face protection in good repair

☐ Self-contained breathing apparatus used in oxygen-deficient environments (less than 19.5 percent oxygen)

(Continued)

FIGURE 8.4 *(Continued)*

☐ Outside spotter present and rescue equipment available at all times when employees are working in confined or enclosed areas where they could be overcome by toxic fumes

☐ Respirators used when working with substances containing toxic vapors, fumes, or dust (only NIOSH/MSHA respirators approved for the specific work conditions)

☐ Disposable respirators used by multiple persons are cleaned before each use

☐ Gloves and eye protection worn by workers welding or working with metal or sharp objects

☐ Appropriate respirators worn during painting or when working with hazardous chemicals

☐ Workers adequately protected overall

Electrical

☐ All equipment either grounded or double-insulated

☐ Power circuits marked with hazard warnings where accidental contact by tools or equipment is a risk

☐ Effective grounding in place of non-current-carrying metal parts for portable or plug-connected equipment

☐ GFCIs installed on all 110-120 V temporary circuits

☐ Temporary lights equipped with guards to prevent accidental contact with bulbs

☐ Receptacles (attachment plugs) not interchangeable with circuits of different voltage

☐ Electrical cords not used for hoisting or carrying tools or equipment

☐ Electrical equipment tested daily

☐ Number of outlets available appropriate for the number of tools in use

☐ Circuit breaker panel clearly labeled and secured

☐ Electrical outlets provided with faceplates

☐ Electrical panel has at least 4 square feet unobstructed clearance

☐ Work areas kept free of cords and excess equipment

☐ Reverse polarity has been checked

Scaffolding

☐ All open sides and ends of platforms more than 10 feet above ground on floor level are provided with top rails, midrails, and toeboards

☐ Top rails 42 inches high ± 3 inches; midrails located midway between floor surface and top rail

☐ Guardrails capable of withstanding 200 pounds of force anywhere along the top rail

☐ Platform planks laid together tightly, preventing tools and material from falling through

☐ Ladders used to gain access to scaffold work platforms

☐ All scaffolding erected per manufacturer's instructions, meets guidelines outlined in OSHA standards

☐ Footing and anchor of scaffolding sound, rigid, and capable of carrying 4 times the maximum intended load without settling or displacement

☐ Base of scaffold on the ground supported by appropriate mudsills

☐ Scaffolding or planking properly supported on scaffold jacks (not on stacks of wood, boxes, bricks, blocks, barrels or other unstable materials)

☐ Lifelines attached to structure and safety belts used on suspended scaffolds

☐ Scaffold planks certified (no substitute materials)

☐ Planks overlap the end of the scaffold no less than 6 inches and no more than 12 inches

☐ Working surface of scaffold fully planked

☐ Scaffold tied off every 30 feet horizontally and 26 feet vertically, or tied off if the height exceeds 4 times the width of the base

Stairways and Ladders

☐ Stairs, ladders, or properly designed ramps provided in access/egress areas that have a change in elevation of more than 19 inches

☐ Handrails provided on stairs with 4 or more steps

☐ Handrails between 30 inches and 37 inches in height

☐ Ladders at risk for displacement tied off and secured

☐ Ladder extends at least 3 feet above the landing

☐ All job-built ladders constructed in accordance with 29 CFR 1926.25 regulations

☐ No metal ladders used within 10 feet of electrical lines

☐ Stepladders used only in full open position

☐ All manufactured single and extension ladders equipped with ladder shoes

☐ All portable ladder rungs have vertical spacing between 10 inches and 14 inches

☐ Extension ladders placed so that the distance from the top support to the ladder base is 1/4 the working distance of the ladder

☐ Faulty ladders tagged "do not use," blocked with plywood, or rendered inoperable

Trenching and Excavation

☐ Walls and faces of excavations where employees are exposed to danger from moving ground guarded by shoring system or by sloping or benching of ground

☐ Type of soil considered in design of slope, benching, or shoring of excavation

(Continued)

FIGURE 8.4 *(Continued)*

☐ Excavated or other material placed a minimum of 2 feet from the edge of excavation

☐ Trenches 5 feet deep or deeper shored or have sloping sides

☐ Trenches more than 4 feet deep have adequate means of exit within 25 feet of travel

☐ All excavations surrounded by barricades

☐ All parts of shoring system in good repair

☐ Excavations no deeper than 2 feet below the base of any shoring system

☐ Equipment manufactured after 1972 has a roll cage

☐ All equipment has operational alarm system when it is in reverse

Personal Fall Arrest System

☐ Guardrail, safety net, or personal fall arrest system used when working more than 6 feet above the ground (exception made for workers installing trusses)

☐ 100 percent tie off required when working at or above 25 feet; tie off requires the use of two lanyards, lifelines, or static lines

☐ Personal fall arrest systems in use adjusted to prevent the user from falling more than 6 feet, or from contacting any lower level

☐ Personal fall arrest systems inspected before each use by a competent person

☐ Lanyard, harness, D-rings, and other personal fall arrest systems in good condition and suitable for use

☐ Fall arrest systems anchored to an appropriate anchorage point capable of withstanding 5,000 pounds of force

☐ Roofing slide guard requirements reflect pitch of roof

Record Keeping

A number of records relating to safety must be kept. Some of these records are required by law, others by good management. OSHA requires you to keep a record of all work-related illnesses, including respiratory problems, and illnesses that result from exposure to toxic substances. OSHA also requires you to keep records of all work-related accidents (other than basic first aid). Specifically, records are required for all injuries that have the following consequences:

- death
- loss of consciousness
- restriction of work or motion
- one or more lost work days
- transfer to another job
- medical treatment other than first aid

OSHA uses specific definitions to distinguish *medical treatment* and *first aid* (Figure 8.5). The superintendent should be aware of these definitions when recording actions taken in response to a work-related injury or illness.

Two types of records must be kept. The first report is called the Employer's First Report of Injury or Illness (OSHA Form 101—see Figure 8.6). This report contains specific, detailed information concerning each work-related illness or injury. All superintendents should see to it that OSHA Form 101 is completed, if possible by the ill or injured employee. The form must be completed within six days of the illness or injury. A copy of this report usually goes to the builder's workers' compensation carrier to assist in processing a claim.

The second report is sometimes called the supplementary report of injuries and illnesses. The Log and Summary of Occupational Injuries and Illnesses (OSHA FORM 200), this report summarizes all illnesses and recordable accidents for a full year (see Figure 8.7). In essence, OSHA FORM 200 digests the information reported on the various copies of OSHA 100.

FIGURE 8.5 Definitions of Medical Treatment and First Aid

OSHA distinguishes medical treatment from first aid. When logging actions taken on the jobsite in response to an illness or accidental injury, be sure to describe the actions carefully and in accordance with the following categorization:

Medical Treatment

Treatment of :
- Infections
- Second- or third-degree burns

Application of:
- Sutures (stitches)
- Butterfly adhesive dressing(s) or strip(s)
- Hot or cold compress(es) during second or subsequent visit to medical personnel
- Heat therapy during second or subsequent visit to medical personnel

Use of:
- Prescription medications (except a single dose administered on first visit for minor injury or discomfort)
- Hot or cold soaking therapy during second or subsequent visit to medical personnel
- Whirlpool bath therapy during second or subsequent visit to medical personnel

Removal of:
- Foreign bodies embedded in eye
- Foreign bodies from wound (if procedure is complicated because of minor injury or discomfort)
- Positive *x*-ray diagnosis of fractures, broken bones, and so forth
- Admission to a hospital or equivalent medical facility for treatment

First Aid Treatment

Application of:
- Antiseptics during first visit to medical personnel
- Hot or cold compress(es) during first visit to medical personnel
- Ointments to abrasions to prevent drying or cracking
- Heat therapy during first visit to medical personnel

Use of:
- Elastic bandage(s) during first visit to medical personnel
- Nonprescription medications and administration of single dose of prescription medication on first visit for minor injury or discomfort
- Whirlpool bath therapy during first visit to medical personnel

Removal of:
- Foreign bodies not embedded in eye (if only irrigation is required)
- Foreign bodies from wound, if procedure is uncomplicated (for example, by using tweezers or by a comparably simple technique)

Treatment of:
- First-degree burn(s)
- Soaking therapy on initial visit to medical personnel or removal of bandages by soaking
- Negative *x*-ray diagnosis
- Observation of injury during visit to medical personnel

Among its other purposes OSHA FORM 200 is intended to keep employees informed concerning the occupational illnesses and injuries that occur at the company. The report is required to be posted in a location where employees have access to the information. The precise location of the posting will vary depending on the size and type of operation of the building company. A large-volume development builder with a fixed site or jobsite trailer may post the report at the jobsite. A scattered-site builder or a small-volume builder whose employees report for work at a company office might post the report at the office. The completed summary record for the previous year must be posted during the month of February each year and placed where other notices to employees are commonly placed. All OSHA records should be maintained on file for at least five years and should be available to employees for inspection.

FIGURE 8.6 OSHA Form 101

OCCUPATIONAL SAFETY AND HEALTH   OSHA FORM NO. 101

REGULATIONS AND PROCEDURES

OSHA No. 101   Form approved
Case or File No. _____   OMB No. 44R 1453

Supplementary Record of Occupational Injuries and Illnesses

EMPLOYER

1. Name _____

2. Mail address _____
            (No. and street)   (City or town)   (State)

3. Location, if different from mail address _____

INJURED OR ILL EMPLOYEE

4. Name _____ Social Security No. _____
        (First name)   (Middle name)   (Last name)

5. Home address _____
          (No. and street)   (City or town)   (State)

6. Age _____   7. Sex: Male _____ Female _____ (Check one)

8. Occupation _____
         (Enter regular job title, *not* the specific activity he was performing at time of injury.)

9. Department _____
         (Enter name of department or division in which the injured person is regularly employed, even
         though he may have been temporarily working in another department at the time of injury.)

THE ACCIDENT OR EXPOSURE TO OCCUPATIONAL ILLNESS

10. Place of accident or exposure _____
         (No. and street)   (City or town)   (State)

    If accident or exposure occurred on employer's premises, give address of plant or establishment in which
it occurred. Do not indicate department or division within the plant or establishment. If accident occurred outside employer's premises at an identifiable address, give that address. If it occurred on a public highway or at any other place which cannot be identified by number and street, please provide place
references locating the place of injury as accurately as possible.

11. Was place of accident or exposure on employer's premises? _____ (Yes or No)

12. What was the employee doing when injured? _____
                      (Be specific. If he was using tools or equipment or handling material,

_____
         name them and tell what he was doing with them.)

_____

13. How did the accident occur? _____
         (Describe fully the events which resulted in the injury or occupational illness. Tell what

_____
happened and how it happened. Name any objects or substances involved and tell how they were involved. Give

_____
full details on all factors which led or contributed to the accident. Use separate sheet for additional space.)

OCCUPATIONAL INJURY OR OCCUPATIONAL ILLNESS

14. Describe the injury or illness in detail and indicate the part of body affected. _____
                      (e.g.: amputation of right index finger

_____
         at second joint; fracture of ribs; lead poisoning; dermatitis of left hand, etc.)

15. Name the object or substance which directly injured the employee. (For example, the machine or thing
he struck against or which struck him; the vapor or poison he inhaled or swallowed; the chemical or radiation which irritated his skin; or in cases of strains, hernias, etc., the thing he was lifting, pulling, etc.)

_____

_____

16. Date of injury or initial diagnosis of occupational illness _____
                      (Date)

17. Did employee die? _____ (Yes or No)

OTHER

18. Name and address of physician _____

19. If hospitalized, name and address of hospital _____

_____

    Date of report _____ Prepared by _____

    Official position _____

Reporting Requirements

Within eight hours after the occurrence of an accident that results in one or more employee fatalities or that results in hospitalization of three or more employees, the employer is required to file a report on accident either orally or in writing to the nearest area director of OSHA.

Hazard Communication Guidelines. While on the job every worker has a right to be protected from obvious hazards and illness. However, building companies and their superintendents now also face the task of informing their workers regarding certain job-related hazards that may not be quite so obvious. OSHA currently has full and immediate authority to conduct hazard communication (HazCom) inspections of all construction companies and jobsites. Citations and possible fines of up to $7,000 per violation have been issued for noncompliance with the HazCom standard.

All building companies and contractors, no matter their size, must comply with the OSHA HazCom standard. Employers must inform and train their employees about all hazardous substances they are working with, as well as about hazardous materials they might come into contact with from other trades.

HazCom compliance has four main elements:

- preparing a written hazard communication program for your company
- labeling products and containers
- providing Material Safety Data Sheets (MSDSs)
- training employees

The National Association of Home Builders, Associated Builders and Contractors, and the American Trade Contractors Association have developed a comprehensive, step-by-step hazard communication compliance kit that has been praised by OSHA. This *Hazard Communication Kit* is available from NAHB and many additional materials about safety and health are available through NAHB's Labor, Safety, and Health Department (see Additional Resources).

Accident Investigation

Superintendents should be trained in a formal accident investigation process. Accidents should be reported to management as soon as possible. Complete an official accident report as promptly as possible following the accident (see Figure 8.6). All accidents should be investigated thoroughly, either by the superintendent or by another manager.

The purpose of an accident investigation is to determine the facts about the accident—not to find fault. Make every effort to be sure you clearly understand what happened so you can prevent similar accidents in the future. Take photographs to document the accident site. Make sketches of the accident scene if necessary. Having certain items available on the jobsite (or easily obtainable from the office) can help you conduct a prompt, orderly investigation in the event of an accident. These items include:

- camera and film
- tape recorder
- tape measure
- sample bottles, tags, and adhesive tape

FIGURE 8.7   OSHA Form 200

Bureau of Labor Statistics
Log and Summary of Occupational
Injuries and Illnesses

NOTE:	This form is required by Public Law 91-596 and must be kept in the establishment for *5 years*. Failure to maintain and post can result in the issuance of citations and assessment of penalties. *(See posting requirements on the other side of form.)*	**RECORDABLE CASES:** You are required to record information about every occupational **death**; every nonfatal occupational **illness**; and those nonfatal occupational **injuries** which involve one or more of the following: loss of consciousness, restriction of work or motion, transfer to another job, or medical treatment (other than first aid). *(See definitions on the other side of form.)*

Case or File Number	Date of Injury or Onset of Illness	Employee's Name	Occupation	Department	Description of Injury or Illness
Enter a nonduplicating number which will facilitate comparisons with supplementary records.	Enter Mo./day.	Enter first name or initial, middle initial, last name.	Enter regular job title, not activity employee was performing when injured or at onset of illness. In the absence of a formal title, enter a brief description of the employee's duties.	Enter department in which the employee is regularly employed or a description of normal workplace to which employee is assigned, even though temporarily working in another department at the time of injury or illness.	Enter a brief description of the injury or illness and indicate the part or parts of body affected. Typical entries for this column might be: Amputation of 1st joint right forefinger; Strain of lower back; Contact dermatitis on both hands; Electrocution--body.
(A)	(B)	(C)	(D)	(E)	(F)
					PREVIOUS PAGE TOTALS ⟶
					TOTALS (Instructions on other side of form.) ⟶

OSHA No. 200   ★U.S.GPO:1990-262-256/15419   FOLD

GURE 8.7 *(Continued)*

U.S. Department of Labor

For Calendar Year 19 _____   Page ____ of ____

Name _____

ment Name _____

ment Address _____

Form Approved
O.M.B. No. 1220-0029
See OMB Disclosure
Statement on reverse.

	and Outcome of INJURY					Type, Extent of, and Outcome of ILLNESS												
	Nonfatal Injuries					Type of Illness								Fatalities	Nonfatal Illnesses			
	Injuries With Lost Workdays				Injuries Without Lost Workdays	CHECK Only One Column for Each Illness *(See other side of form for terminations or permanent transfers.)*								Illness Related	Illnesses With Lost Workdays				Illnesses Without Lost Workdays
TE Enter a CHECK if injury involves days away from work, or days of restricted work activity, or both.	Enter a CHECK if injury involves days away from work.	Enter number of DAYS *away from work.*	Enter number of DAYS of *restricted work activity.*	Enter a CHECK if no entry was made in columns 1 or 2 but the injury is recordable as defined above.	Occupational skin diseases or disorders	Dust diseases of the lungs	Respiratory conditions due to toxic agents	Poisoning (systemic effects of toxic materials)	Disorders due to physical agents	Disorders associated with repeated trauma	All other occupational illnesses	Enter DATE of death. Mo./day/yr.	Enter a CHECK if illness involves days away from work, or days of restricted work activity, or both.	Enter a CHECK if illness involves days away from work.	Enter number of DAYS *away from work.*	Enter number of DAYS of *restricted work activity.*	Enter a CHECK if no entry was made in columns 8 or 9.
yr. (2)	(3)	(4)	(5)	(6)	(a)	(b)	(c)	(d)	(e)	(f)	(g)						
											(7)	(8)	(9)	(10)	(11)	(12)	(13)

(blank data rows)

cation of Annual Summary Totals By _____ Title _____ Date _____

No. 200   **POST ONLY THIS PORTION OF THE LAST PAGE NO LATER THAN FEBRUARY 1.**

(Continued)

FIGURE 8.7 (Continued)

OMB DISCLOSURE STATEMENT

We estimate that it will take from 4 minutes to 30 minutes to complete a line entry on this form, including time for reviewing instructions; searching, gathering and maintaining the data needed; and completing and reviewing the entry. If you have any comments regarding this estimate or any other aspect of this recordkeeping system, send them to the Bureau of Labor Statistics, Division of Management Systems (1220-0029), Washington, D.C. 20212 and to the Office of Management and Budget, Paperwork Reduction Project (1220-0029), Washington, D.C. 20503.

Instructions for OSHA No. 200

I. Log and Summary of Occupational Injuries and Illnesses

Each employer who is subject to the recordkeeping requirements of the Occupational Safety and Health Act of 1970 must maintain for each establishment a log of all recordable occupational injuries and illnesses. This form (OSHA No. 200) may be used for that purpose. A substitute for the OSHA No. 200 is acceptable if it is as detailed, easily readable, and understandable as the OSHA No. 200.

Enter each recordable case on the log within six (6) workdays after learning of its occurrence. Although other records must be maintained at the establishment to which they refer, it is possible to prepare and maintain the log at another location, using data processing equipment if desired. If the log is prepared elsewhere, a copy updated to within 45 calendar days must be present at all times in the establishment.

Logs must be maintained and retained for five (5) years following the end of the calendar year to which they relate. Logs must be available (normally at the establishment) for inspection and copying by representatives of the Department of Labor, or the Department of Health and Human Services, or States accorded jurisdiction under the Act. Access to the log is also provided to employees, former employees and their representatives.

II. Changes in Extent of or Outcome of Injury or Illness

If, during the 5-year period the log must be retained, there is a change in an extent and outcome of an injury or illness which affects entries in columns 1, 2, 6, 8, 9, or 13, the first entry should be lined out and a new entry made. For example, if an injured employee at first required only medical treatment but later lost workdays away from work, the check in column 6 should be lined out, and checks entered in columns 2 and 3 and the number of lost workdays entered in column 4.

In another example, if an employee with an occupational illness lost workdays, returned to work, and then died of the illness, any entries in columns 9 through 12 should be lined out and the date of death entered in column 8.

The entire entry for an injury or illness should be lined out if later found to be nonrecordable. For example: an injury which is later determined not to be work related, or which was initially thought to involve medical treatment but later was determined to have involved only first aid.

III. Posting Requirements

A copy of the totals and information following the fold line of the last page for the year must be posted at each establishment in the place or places where notices to employees are customarily posted. This copy must be posted no later than *February 1 and must remain in place until March 1.*

Even though there were no injuries or illnesses during the year, zeros must be entered on the totals line, and the form posted.

The person responsible for the *annual summary totals* shall certify that the totals are true and complete by signing at the bottom of the form.

IV. Instructions for Completing Log and Summary of Occupational Injuries and Illnesses

Column A — CASE OR FILE NUMBER. Self-explanatory.

Column B — DATE OF INJURY OR ONSET OF ILLNESS.
For occupational injuries, enter the date of the work accident which resulted in injury. For occupational illnesses, enter the date of initial diagnosis of illness, or, if absence from work occurred before diagnosis, enter the first day of the absence attributable to the illness which was later diagnosed or recognized.

Columns
C through F — Self-explanatory.

Columns
1 and 8 — INJURY OR ILLNESS-RELATED DEATHS.
Self-explanatory.

Columns
2 and 9 — INJURIES OR ILLNESSES WITH LOST WORKDAYS.
Self-explanatory.

Any injury which involves days away from work, or days of restricted work activity, or both must be recorded since it always involves one or more of the criteria for recordability.

Columns
3 and 10 — INJURIES OR ILLNESSES INVOLVING DAYS AWAY FROM WORK. Self-explanatory.

Columns
4 and 11 — LOST WORKDAYS—DAYS AWAY FROM WORK.
Enter the number of workdays (consecutive or not) on which the employee would have worked but could not because of occupational injury or illness. The number of lost workdays should not include the day of injury or onset of illness or any days on which the employee would not have worked even though able to work.
NOTE: For employees not having a regularly scheduled shift, such as certain truck drivers, construction workers, farm labor, casual labor, part-time employees, etc., it may be necessary to estimate the number of lost workdays. Estimates of lost workdays shall be based on prior work history of the employee AND days worked by employees, not ill or injured, working in the department and/or occupation of the ill or injured employee.

Columns
5 and 12 — LOST WORKDAYS—DAYS OF RESTRICTED WORK ACTIVITY.
Enter the number of workdays (consecutive or not) on which because of injury or illness:
(1) the employee was assigned to another job on a temporary basis, or
(2) the employee worked at a permanent job less than full time, or
(3) the employee worked at a permanently assigned job but could not perform all duties normally connected with it.

The number of lost workdays should not include the day of injury or onset of illness or any days on which the employee would not have worked even though able to work.

FIGURE 8.7 *(Continued)*

Columns
6 and 13 — INJURIES OR ILLNESSES WITHOUT LOST
WORKDAYS. Self-explanatory.

Columns 7a
through 7g — TYPE OF ILLNESS.
Enter a check in only *one* column for each illness.

TERMINATION OR PERMANENT TRANSFER—Place an asterisk to the right of the entry in columns 7a through 7g (type of illness) which represented a termination of employment or permanent transfer.

V. Totals

Add number of entries in columns 1 and 8.
Add number of checks in columns 2, 3, 6, 7, 9, 10, and 13.
Add number of days in columns 4, 5, 11, and 12.
Yearly totals for each column (1-13) are required for posting. Running or page totals may be generated at the discretion of the employer.

If an employee's loss of workdays is continuing at the time the totals are summarized, estimate the number of future workdays the employee will lose and add that estimate to the workdays already lost and include this figure in the annual totals. No further entries are to be made with respect to such cases in the next year's log.

VI. Definitions

OCCUPATIONAL INJURY is any injury such as a cut, fracture, sprain, amputation, etc., which results from a work accident or from an exposure involving a single incident in the work environment.
NOTE: Conditions resulting from animal bites, such as insect or snake bites or from one-time exposure to chemicals, are considered to be injuries.

OCCUPATIONAL ILLNESS of an employee is any abnormal condition or disorder, other than one resulting from an occupational injury, caused by exposure to environmental factors associated with employment. It includes acute and chronic illnesses or diseases which may be caused by inhalation, absorption, ingestion, or direct contact.

The following listing gives the categories of occupational illnesses and disorders that will be utilized for the purpose of classifying recordable illnesses. For purposes of information, examples of each category are given. These are typical examples, however, and are not to be considered the complete listing of the types of illnesses and disorders that are to be counted under each category.

7a. **Occupational Skin Diseases or Disorders**
Examples: Contact dermatitis, eczema, or rash caused by primary irritants and sensitizers or poisonous plants; oil acne; chrome ulcers; chemical burns or inflammations; etc.

7b. **Dust Diseases of the Lungs (Pneumoconioses)**
Examples: Silicosis, asbestosis and other asbestos-related diseases, coal worker's pneumoconiosis, byssinosis, siderosis, and other pneumoconioses.

7c. **Respiratory Conditions Due to Toxic Agents**
Examples: Pneumonitis, pharyngitis, rhinitis or acute congestion due to chemicals, dusts, gases, or fumes; farmer's lung; etc.

7d. **Poisoning (Systemic Effect of Toxic Materials)**
Examples: Poisoning by lead, mercury, cadmium, arsenic, or other metals; poisoning by carbon monoxide, hydrogen sulfide, or other gases; poisoning by benzol, carbon tetrachloride, or other organic solvents; poisoning by insecticide sprays such as parathion, lead arsenate; poisoning by other chemicals such as formaldehyde, plastics, and resins; etc.

7e. **Disorders Due to Physical Agents (Other than Toxic Materials)**
Examples: Heatstroke, sunstroke, heat exhaustion, and other effects of environmental heat; freezing, frostbite, and effects of exposure to low temperatures; caisson disease; effects of ionizing radiation (isotopes, X-rays, radium); effects of nonionizing radiation (welding flash, ultraviolet rays, microwaves, sunburn); etc.

7f. **Disorders Associated With Repeated Trauma**
Examples: Noise-induced hearing loss; synovitis, tenosynovitis, and bursitis; Raynaud's phenomena; and other conditions due to repeated motion, vibration, or pressure.

7g. **All Other Occupational Illnesses**
Examples: Anthrax, brucellosis, infectious hepatitis, malignant and benign tumors, food poisoning, histoplasmosis, coccidioidomycosis, etc.

MEDICAL TREATMENT includes treatment (other than first aid) administered by a physician or by registered professional personnel under the standing orders of a physician. Medical treatment does NOT include first-aid treatment (one-time treatment and subsequent observation of minor scratches, cuts, burns, splinters, and so forth, which do not ordinarily require medical care) even though provided by a physician or registered professional personnel.

ESTABLISHMENT: A single physical location where business is conducted or where services or industrial operations are performed (for example: a factory, mill, store, hotel, restaurant, movie theater, farm, ranch, bank, sales office, warehouse, or central administrative office). Where distinctly separate activities are performed at a single physical location, such as construction activities operated from the same physical location as a lumber yard, each activity shall be treated as a separate establishment.

For firms engaged in activities which may be physically dispersed, such as agriculture; construction; transportation; communications; and electric, gas, and sanitary services, records may be maintained at a place to which employees report each day.

Records for personnel who do not primarily report or work at a single establishment, such as traveling salesmen, technicians, engineers, etc., shall be maintained at the location from which they are paid or the base from which personnel operate to carry out their activities.

WORK ENVIRONMENT is comprised of the physical location, equipment, materials processed or used, and the kinds of operations performed in the course of an employee's work, whether on or off the employer's premises.

- cord and barriers tape
- accident report forms
- First Aid kit (on the jobsite as part of safety program)
- carrying bags
- portable lamp or flashlight
- special equipment such as meters, detectors, test equipment

Of course, the first priorities when an accident occurs are to control the situation, get help for the accident victim, and—if the hazard is ongoing—eliminate the danger to prevent further injury. Note that "control the situation" comes first. An injured worker will benefit most from calm, orderly assistance and may even suffer further harm if given well-meaning but inappropriate aid. Other workers on the jobsite are likely shocked, confused, and frightened. Some workers may be anxious to help, particularly if they are friends with the injured person. Depending on the specific situation, the superintendent may delegate certain tasks. By promptly taking charge of the situation, however, the superintendent accomplishes the following things:

- ensures that the injured worker receives prompt, appropriate help
- prevents further injuries, particularly in a situation of ongoing danger
- alleviates the anxieties of the other workers on the site
- reduces disruption of the work

The superintendent must also be sure to notify management of the incident and secure the area (both to make it safe and to ensure the accuracy of the accident investigation). When the immediate danger has been contained and any injured workers have received help, the investigation can begin. A thorough investigation involves many steps and is designed to document as many details about the incident as possible (see Figure 8.8).

After interviewing all witnesses, the investigator should analyze each witness's statement. You may wish to re-interview one or more witnesses to confirm or clarify key points. While there may be inconsistencies in witnesses' statements, the investigator should assemble the available testimony into a logical order. Analyze this information along with data from the accident site.

Not all people react in the same manner to a particular stimulus. For example, a witness within close proximity to the accident may present an entirely different story from one who saw it at a distance. Some witnesses may also change their stories after they have discussed the accident with other people. The reason for the change may (or may not) be additional clues.

A witness who has had a traumatic experience may not be able to recall the details of the accident. A witness who has a vested interest in the results of the investigation may offer biased testimony. Finally, eyesight, hearing, reaction time, and the general condition of each witness may affect his or her powers of observation. A witness may omit entire sequences because of a failure to observe them or because their importance was not realized.

Trade Contractor Safety

Trade contractors should develop their own safety and health programs. Each trade contractor is responsible for the safety of his or her employees. Meanwhile, however, the trade contractor is on your jobsite. Therefore, the rules for trade contractors should be similar to the rules for your employees. All trade contractors and their

FIGURE 8.8 Accident Investigation

Step 1. Control the Situation

- Send for help
- Administer First Aid
- Notify management
- Secure the area and make it safe

Step 2. Stop the Ongoing Hazard

Depending on the situation, one or more of the following actions may be necessary:

- Shut off electricity
- Bleed or isolate pressurized systems
- Block mechanical equipment
- Check air quality
- Issue personal protective equipment
- Provide emergency lighting, power, air, and so forth
- Secure the accident scene from unauthorized access and protect the evidence
- Rope off the area or station a guard
- Issue lockout, tag-out permits

Step 3. Collect Evidence

- Take notes
- Take lots of photographs (long-distance and close-up shots; sign and date Polaroids)
- Check position and condition of tools, layout, and so forth
- Note air quality, things that evaporate or melt, and so forth
- Note weather and lighting conditions
- Note the presence and locations of tire tracks, foot prints, loose material on the floor, and so forth
- Collect operating logs, charts, records, and so forth
- Note identification numbers on equipment and check maintenance records
- Make sketches of the accident site, note dimensions
- Review the general conditions of the jobsite at the time of the accident, looking for contributing factors such as the following:
  - Housekeeping
  - Periodic rule or procedure violations (check log records and personnel files)
  - Work environment
  - Training, experience, or supervision
  - Floor or surface conditions
  - Alcohol or drug abuse
  - Lighting or visibility
  - Employee morale or attitude
  - Noise or distractions
  - Health or safety record
  - Air quality, treatment, or weather
  - Equipment condition or malfunction history

Step 4. Get Things Back to Normal as Soon as Practicable

Step 5. Interview Witnesses

- Establish an order in which to obtain comments from witnesses
- Obtain preliminary statements from all witnesses as soon as possible
- Locate the position of each witness on a master chart (including the direction of view)
- Arrange for a convenient time and place to talk further with each witness
- Explain the purpose of the investigation (accident prevention) and put each witness at ease
- Listen, let each witness speak freely and be courteous and considerate
- Take notes without distracting the witness (Use a tape recorder only with the consent of the witness.)
- Use sketches and diagrams to help the witness
- Emphasize areas of direct observation. Label hearsay accordingly in your notes
- Be sincere and do not argue with the witness
- Record the exact words used by the witness to describe each observation; do not "put words into the witness's mouth"
- Word each question carefully; be sure the witness understands each question
- Identify the qualifications of each witness (name, address, occupation, years of experience, and so forth)
- Supply each witness with a copy of his or her statement (If possible, have each witness sign his or her statement.)

Step 6. Analysis

- Write down the accident story
- Note the undisputed facts and the disputed facts
- Compare the disputed facts, the undisputed facts, and the physical evidence
- Compete the accident report by recording the following information:
  - Who had the accident?

(Continued)

FIGURE 8.8 *(Continued)*

— What happened?
— Where did it occur?
— When did it happen?
— When was it reported?
— Why did it happen?
— What caused it?
— What contributed to it?
— What was the direct cause?
— What was the indirect cause or causes?
— What was the root causes or causes?

— Who had the most control over the situation?
— Who witnessed the accident? (Record names, addresses, telephone numbers)
— What can be done to prevent this kind of accident in the future?
— Who will be responsible to make sure it doesn't happen again?
— What changes to the safety program or operational procedures need to be made?
— When will the changes be made?

employees are required to comply with the builder's jobsite safety and drug policies, and all the Occupations Safety and Health Construction Standards along with other state and local safety requirements while working on any project.

As a condition of their contract with the builder, trade contractors should be willing to do the following:

- comply with the applicable federal and state OSHA regulations
- supply the builder with a copy of the trade contractor's company safety program and material safety data sheets (MSDSs) for materials used on the builder's projects. (If a trade contractor does not have a safety program, most builders require the trade contractor to comply with the builder's safety program.)
- report any accident, injury, or fatality immediately to the superintendent
- supply the proper personal protective equipment and safety equipment to the trade contractors' employees and assure their use
- adequately train their field employees on proper safety practices
- report unsafe conditions to the superintendent
- notify the superintendent (or the builder's office when no superintendent is onsite) immediately in the event of an OSHA inspection

Trade contractors may be subject to disciplinary action for the following actions:

- failure to perform work
- inefficient performance, incompetence, or neglect of work
- willful refusal to work as directed (insubordination)
- negligence in observing safety regulations, poor housekeeping, or failure to report on-the-job injuries or unsafe conditions
- unexcused or excessive delays in the work
- discourtesy, irritations, friction, conduct creating disharmony, or unwillingness or inability to work in harmony with others
- horseplay, fighting, threatening, intimidating or coercing others on the jobsite or company premises
- bringing unauthorized weapons, firearms, or explosives on the jobsite or in their vehicles
- harassing or discriminating against an individual
- failure to be prepared for work by wearing the appropriate construction clothing or bringing the necessary tools
- violation of any other commonly accepted reasonable rule of responsible personal conduct

Some offenses are intolerable and must be dealt with sternly whether the offense is committed by the builder's employee or by a trade contractor's employee. Intolerable offenses or actions may include:

- dishonesty or falsification in any form or to any degree
- damage, loss or destruction of company, employee, or customer property due to willful or negligent acts
- unauthorized possession, removal, or use of property belonging to the builder, customers, employees, or other trade contractors.
- being under the influence of or in possession of alcoholic beverages, intoxicants, or illegal drugs on the job where the influence of alcohol is .02 or greater
- refusal to submit to a drug screening or failure to pass a drug screening
- safety violations that endanger employees or other trade contractors

Enforcement and Discipline

If a trade contractor repeatedly violates safety standards or his or her response to a request for compliance is not accomplished within the specified period of time, the superintendent may invoke a *stop work order* until the violations are corrected. Chronic or serious violators should not be used on further projects. For minor offenses with minor consequences, a trade contractor will be expected to agree to improve behavior. Minor offenses may be recorded as a warning. Suspension of a trade contractor's contract or termination of the contract should result from major offenses, those with serious or costly consequences, or for repeated minor offenses when the trade contractor shows a lack of responsible effort to correct deficiencies.

The builder's safety policies must be enforced fairly and diligently. Disciplinary action applies to employees as well as trade contractors. A three-step approach to discipline is recommended:

1. Discuss a first infraction with the employee or trade contractor. Recap the discussion in a written reprimand to document the infraction and the fact that you have addressed it with the employee or trade contractor.

2. With a second infraction, especially if it is a repeat of the first offense, give an employee time off without pay. (This level of disciplinary action is comparable to issuing a stop work order for a trade contractor.)

3. A third violation results in dismissal of the employee. For a trade contractor, three violations would result in dismissal from the job and cancellation of any future work with the builder.

The Emergency Action Plan

Each builder should develop an emergency action plan designed to help protect the health and safety of workers and to provide the best possible emergency care for employees, trade contractors, and visitors to the jobsite.

A builder's emergency action plan includes emergency telephone numbers for medical, fire, police, hazardous materials response unit, Environmental Protection Agency (EPA), OSHA, the project superintendent, and the builder's office. It should also contain policies and procedures for medical emergencies, fire emergencies, and property or equipment damage in the event of an accident.

The following guidelines should be included in the emergency action plan:

- definition of an emergency
- evacuation criteria and guidelines
- First Aid and CPR requirements and treatment guidelines
- medical assistance guidelines for life-threatening accidents
- medical assistance guidelines for non-life-threatening accidents
- cleanup and disposal guidelines for hazardous spills
- accident investigation and reporting guidelines
- notification requirements for the company and for various agencies

If an evacuation is necessary, the superintendent will select an appropriate site for evacuation. Make certain everyone is properly evacuated and accounted for.

The superintendent—or some other competent person on site—should know how to perform first aid, including CPR. Basic First Aid can be administered on site for very minor injuries. If a life-threatening injury occurs the superintendent should contact the nearest emergency medical facility. If the superintendent is not on the job-site at the time of the incident, the crew leader or other responsible party should contact the emergency facility. If the injury is not life threatening, the superintendent should arrange for the accident victim to be taken to the company-preferred medical provider.

In addition to notifying the medical provider, the company office should be contacted.

Also, for certain accidents involving fatalities or multiple serious injuries OSHA must be notified within eight hours.

Teamwork

Safety is a team effort. Employees or management alone cannot accomplish the objective of zero injuries and a safe working environment. It takes a concentrated effort from top management, coupled with the dedication from superintendents and commitment from employees and trade contractors to make safety a priority. It is worth the effort! Money can be saved, productivity improved, and the health and safety of employee can be safeguarded.

Quality Inspection Checklists

Excavation Checklist

Trade Contractor: _____   Job Name, Number: _____

- ☐ Silt fence installed as required
- ☐ Culvert installed as required
- ☐ Strings on property line pulled if there is a question of setback requirements
- ☐ Footings extended down to virgin soil if any existing fill will interfere with the footings
- ☐ Loose soil or rock removed so footings rest on solid ground
- ☐ Stumps and trees disposed of according to instructions
- ☐ Basement floor leveled with a builder's level ± 1/4 inch in 20 feet or 1/2 inch overall
- ☐ Room left for waterproofing around basement foundations; minimum, 2 feet
- ☐ Pathways for water lines or well trenches clearly marked, as required
- ☐ Pathways to septic tank and lines clearly marked, as required
- ☐ Pathways to power line trenches clearly marked, as required
- ☐ Culvert and gravel for driveway installed, as required; driveway solid enough for trucks and other heavy equipment; access to the foundation adequate for the site
- ☐ Silt screen reinstalled following grading as needed
- ☐ Superintendent contacted immediately if ground water is encountered

Superintendent's Signature: _____   Date: _____

Layout and Footing Checklist

Trade Contractor: _____   Job Name, Number: _____

☐ Work complies with local, state, and the national codes (even in areas where no local inspection exists)

Location, Layout

☐ Property lines surveyed by licensed surveyor or certified by the owner

☐ Property survey stakes in place

☐ Plot plan available

☐ Layout confirmed to be per owner's agreement at the site meeting or plat for the development

☐ Correct front yard setback, _____ feet

☐ Parallel to the street to within 1 inch if applicable

☐ Correct side yard setback, _____ feet Which side? _____

☐ Rear yard setback, _____ feet

☐ Groundwater or soft spots in the soil checked for

☐ Footing below the frost line after backfill completed, if applicable

☐ Drainage destination for footing determined

☐ Elevation checked; should provide for 5 percent grade away from building after backfill

Footings

☐ Inspector notified at least 24 hours ahead of time or in accordance with local inspection requirements

☐ Inspection completed and inspection card signed

☐ Footings checked for position and proper grade; layout double-checked

☐ Dirt thrown to the outside

☐ Soil under footing undisturbed or compacted and adequate to support structure

☐ Sides of footings cut square; forms in place and properly braced

☐ Soil unfrozen, no snow or ice

☐ All dimensions checked

☐ Length is correct, all sides, to within $+\frac{1}{2}$ inch (but not short)

☐ Square $\pm \frac{1}{4}$ inch in 20 feet

☐ Diagonal measurements equal $\pm \frac{1}{2}$ inch

☐ Size, location, and elevation of blockouts checked

☐ Offsets and jogs checked; location _____ ; size _____

☐ Footing depth in inches _____ inches ± 1 inch; width in inches _____ inches ± 1 inch, per plans and local code

☐ Footing steps in 8-inch increments

☐ Metal rebar grade stakes 5 feet on center and set to grade with a builder's level

☐ Forms level ± $\frac{1}{4}$ inch in 10 feet and level ± $\frac{1}{2}$ inch overall

☐ Forms properly braced and backfilled

☐ Footings free of roots and topsoil

☐ Footings free of debris

☐ Proper steps taken to keep concrete from freezing or drying too fast

Horizontal Reinforcing Steel (Where Applicable)

☐ Horizontal rebar size $\frac{1}{2}$ inch; typically 2 pieces

☐ Rebar located in bottom third of footing, not on the ground

☐ Steel has concrete covering of at least 3 inches

☐ Laps 15 inches minimum

☐ Rebar free from rust or oil

☐ Rebar located and spaced ± 1 inch from specified

☐ Reinforcing continuous through cold joints

Concrete, at Time of Pour

☐ Footing concrete 2,500 or 3,000 psi as required Check batch ticket to make sure proper mix was batched

☐ Maximum slump 5 inches ± 1inch

☐ One continuous pour, if at all possible

☐ Concrete is deposited as nearly as possible to final position

☐ Concrete not dropped more than 40 inches

☐ Concrete not over-vibrated

☐ Concrete finished flat and level ± $\frac{1}{4}$ inch in 10 feet, level ± $\frac{1}{2}$ inch overall and floated semi-rough

Concrete after Pour

☐ All forms stripped and removed following proper curing time depending on temperature (not less than 12 hours); scrap material placed in designated location

☐ Concrete batch tickets given to superintendent

Superintendent's Signature: _____   Date: _____

Block Foundation Checklist

Trade Contractor: _____   Job Name, Number: _____

☐ Work complies with local, state, and the national codes (even in areas where no local inspection exists)

Dimensions

☐ Location: setbacks and alignment rechecked

☐ Thickness of foundation _____ inches

☐ All dimensions checked

☐ Foundation not less than specified and not more than $\frac{1}{4}$ inch greater in any dimension (height, length, width, thickness)

☐ Overall length _____ feet

☐ Overall width _____ feet

☐ Offsets checked for location and size

☐ Windows checked for location and size

☐ Windows properly braced

☐ Windows and doors placed within ± 1 inch of specified horizontal dimensions and ± $\frac{1}{2}$ inch of specified vertical dimensions

☐ Windows and doors placed level ± $\frac{1}{8}$ inch

☐ Windows and doors sized no less than and to within $\frac{1}{4}$ inch in all dimensions

☐ Doors checked for location and size

☐ Doors properly braced

☐ Walls plumb ± $\frac{1}{4}$ inch in 8 feet (check several places)

☐ Walls square ± $\frac{1}{4}$ inch in 20 feet; not more than 1 inch out of square in any dimension

☐ Walls straight ± $\frac{1}{4}$ inch in 20 feet; not more than ± $\frac{1}{2}$ inch overall in any wall

☐ Walls level ± $\frac{1}{4}$ inch in 10 feet; not more than ± $\frac{1}{2}$ inch overall in any wall

☐ Lintels installed correctly

☐ Crawl space access doors provided at designated location, size, and height

☐ The top of the foundation is at least 8 inches above the level of the final grade, unless a brick ledge is provided

☐ Concrete block foundation is designated height

☐ Pilasters installed according to plans

☐ Concrete block piers placed in garage as specified

☐ All exterior joints that are not to have a stucco finish are concave, struck neatly, and all tags removed

☐ All interior mortar joints in basements are struck and all excess mortar tags removed

☐ All excess mortar cut off flush with the block on the interior of crawl spaces and on the exterior where stucco is to be applied

☐ Wall ties installed on walls that are to receive brick veneer

☐ Extra block removed from foundation, stacked on a pallet, and counted

☐ All excess mortar cleaned off footings and basement concrete slabs

☐ All trash, including mortar bags, placed in dumpster, trash pile, or other designated area; one location only

☐ Foundation bolts or straps spaced not more that 6 feet on center or as required by code

☐ A minimum of 2 bolts or straps per sill; one strap within 16 inches of the end of each plate

☐ Straps placed in line on frame wall ± $\frac{1}{2}$ inch

☐ Foundation bolts not more than $2\frac{1}{2}$ inches above block foundation based on $1\frac{1}{2}$ inch sill plate

☐ Inspector notified at least 24 hours ahead of time or in accordance with local inspection requirements

☐ Inspection completed and inspection card signed

Superintendent's Signature: _____ Date: _____

Poured Concrete Foundations Checklist

Trade Contractor: _____   Job Name, Number: _____

- ☐ Work complies with local, state, and the national codes (even in areas where no local inspection exists)
- ☐ Door and window openings braced vertically 16 inches on center and (if over 2 feet tall) every 24 inches horizontally to prevent bulging
- ☐ Windows and doors nailed to forms, both sides 12 inches on center
- ☐ Bulkheads checked for location and size
- ☐ Blockouts for plumbing, soil pipe, and so forth checked for location and size
- ☐ Adequate snap ties
- ☐ Walls plumb ± ¼ inches in 8 feet (check several places)
- ☐ Walls square ± ¼ inch in 20 feet; not more than 1 inch out of square in any dimension
- ☐ Walls straight ± ¼ inch in 20 feet; not more than ± ½ inch overall in any wall
- ☐ Walls level ± ¼ inch in 10 feet; not more than ± ½ inch overall in any wall
- ☐ Forms clean and free from concrete buildup
- ☐ Forms properly braced and tied
- ☐ Forms tight to prevent any leakage
- ☐ The top of the foundation is at least 8 inches above the level of the final grade unless a brick ledge is provided
- ☐ Check for missing ties, pins, and connectors
- ☐ Walers solid and properly fastened

Vertical Reinforcing Steel

- ☐ Rebar size: _____ #; Grade: _____
- ☐ Spacing as specified _____ ± 1 inch
- ☐ Steel covered with at least 1½ inches of concrete
- ☐ Steel free from rust or oil
- ☐ Steel double-tied or wired at laps

Horizontal Reinforcing Steel

- ☐ Rebar size: _____ #
- ☐ Spacing as specified _____ ± 1 inch
- ☐ Steel covered with at least 1½ inches of concrete
- ☐ Steel double-tied or wired at laps
- ☐ Steel free from rust or oil
- ☐ Steel tied to vertical rebar with wire at each intersection

Foundation Straps or Bolts

☐ Plate layout established before straps or bolts inserted

☐ Foundation straps or bolts placed at 6 foot intervals within the first ½ hour after final placing of concrete

Concrete

☐ Foundation concrete 3,000 to 3,500 psi as required; check batch ticket to verify proper mix was batched

☐ Maximum slump 5 inches ± 1 inch

☐ One continuous pour

☐ Concrete deposited as near as possible to final position

☐ Concrete placed in lifts of 18 inches to 24 inches maximum

☐ Concrete vibrated at 6-foot to 10-foot intervals

☐ Concrete not vibrated for longer than 15 seconds in one place

☐ Concrete finished flat and level ± ¼ inch in 10 feet; not more than ½ inch overall and left semi-smooth

☐ Forms stripped and removed following proper curing time depending on temperature (not less than 12 hours, and typically 12 to 24 hours after pouring, unless sufficient strength has not yet been reached

☐ Concrete protected from freezing by heating or insulation if necessary

☐ Form ties broken off and tie holes filled with mortar or plastic cement after concrete has had time to cure 3 to 5 days

☐ No visible cracks; any voids, cracks, or honeycomb filled

☐ Honeycomb, if present, filled over with mortar or—if it could affect structural integrity—properly repaired or replaced

☐ Waterproofing applied with complete coverage (no missed spots)

Superintendent's Signature: _____ Date: _____

Dampproofing Checklist

Trade Contractor: _____   Job Name, Number: _____

☐ Work complies with local, state, and national codes, including the Model Energy Code (even in areas where no local inspection exists)

☐ Trash and dirt on footings and in trenches cleaned out

☐ Footing and foundation drains and/or gravel properly installed all the way around the basement

☐ Foundation drainpipes (if used) installed on 2-inch gravel bed and covered with minimum 6 inches gravel on top

☐ Stoops and jogs in foundation have drainpipes and/or gravel around them

☐ Foundation exit drain line (if used) has fall of at least $\frac{1}{4}$ inch per foot for at least 20 feet away from the house

☐ Leave 10 feet of extra pipe for final grader when necessary

☐ Foundation drains do not discharge onto neighbors' property

☐ Drainpipe not crushed, clogged, or damaged

☐ Parging applied $\frac{3}{8}$ inch thick all the way down to footings on basements

☐ Parging behind all waterproofing on basements

☐ Asphalt emulsion applied at step downs and bottom edges of crawl space blocks to prevent seepage

☐ Waterproofing applied around entire house on partial basements

Superintendent's Signature: _____   Date: _____

Backfill Checklist

Trade Contractor: _____   Job Name, Number: _____

- [ ] Work complies with local, state, and the national codes (even in areas where no local inspection exists)
- [ ] Foundation voids, cracks, or honeycomb patched and sealed watertight
- [ ] Wall ties in poured concrete foundations broken off and all tie holes filled with black plastic cement
- [ ] Seam between foundation and footing sealed watertight
- [ ] Basement foundation dampproofing completed according to quality standards
- [ ] Footing drains installed properly according to plans where appropriate
- [ ] Debris and garbage removed from trenches, around foundation, and in backfill area
- [ ] French drain in place, if required
- [ ] Utilities and plumbing connections approved and inspected, as required
- [ ] Window wells in place, where required
- [ ] Foundation properly braced
- [ ] Backfill done in 18-inch lifts ± 6 inches and compacted as required
- [ ] Rock-free and debris-free dirt used as fill
- [ ] Swales completed as required to maintain 10 percent slope or 12 inches in the first 10 feet and 5 percent or 6 inches minimum for the next 10 feet
- [ ] Foundation drains are marked with a stake and flagged and left open and clear
- [ ] Backfill surface is at least 8 inches down from the top of the foundation
- [ ] Silt screens reinstalled or repaired as needed

Superintendent's Signature: _____   Date: _____

Exterior Concrete Flatwork Checklist

Trade Contractor: _____   Job Name, Number: _____

Garage Slabs

- ☐ Work complies with local, state, and the national codes (even in areas where no local inspection exists)
- ☐ Front edge of garage floor slab under garage door turned down so concrete is at least 6 inches thick
- ☐ Garage piers all in place per the plans
- ☐ Concrete reinforcing steel installed as required
- ☐ Rebar raised up off the ground at least $1\frac{1}{2}$ inches with chairs, rocks, or other suitable material; rebar does not rest on the ground
- ☐ Vapor barrier in place, if required
- ☐ Concrete mixed at least 20 minutes and not longer than 1 hour prior to placement
- ☐ Concrete slopes from back of the garage toward the front garage door at least $\frac{1}{8}$ inch per foot or as specified
- ☐ Garage door lip slopes $\frac{1}{2}$ inch at the garage door
- ☐ Concrete perfectly level under the garage door
- ☐ No humps or low spots in garage floor greater than $\frac{1}{4}$ inch; concrete will not allow water to puddle
- ☐ Garage walls protected from concrete splatters
- ☐ Concrete 3,000 psi
- ☐ No cold joints in concrete floor
- ☐ Concrete has smooth trowel finish
- ☐ Area cleaned up; no excess concrete left around the site

Superintendent's Signature: _____   Date: _____

Exterior Concrete Flatwork Checklist

Trade Contractor: _____ Job Name, Number: _____

Walks and Drives

☐ Work complies with local, state, and the national codes (even in areas where no local inspection exists)

☐ Slab checked for proper dimensions ± 1 inch and square ± 1 inch

☐ Area free from snow, ice, and debris

☐ Gravel, where required, placed level ± 1 inch and at depth specified ± 1 inch

☐ 6 mil poly vapor barrier in place, if required

☐ Concrete at least 4 inches thick but never less than 3½ inches

☐ Concrete placed using continuous pour with no cold joints

☐ Concrete deposited as near as possible to final position

☐ Driveway is ½ inch below garage slab to prevent water from entering garage

☐ Concrete 3,000 psi

☐ Maximum slump 4 inches ± 1 inch

☐ Air entrainment 6 percent ± 1 percent in areas exposed to freezing and thawing (very important)

☐ Concrete mixed at least 20 minutes and not longer than 1 hour prior to placement

☐ Concrete to proper grade ± ¼ inch in 10 feet and 1 inch overall; concrete slopes at least ¼ inch per foot or 2 percent for proper drainage

☐ Control joints placed in concrete driveways 12 feet on center; no wood expansion joints in driveways

☐ Control joints in sidewalks are 5 feet on center

☐ No water sprinkled on concrete surface to make troweling easier

☐ No premature finishing and steel trowel finishing while bleed water present

☐ Concrete has broom finish or other suitable exterior finish

☐ Water has an escape route from all concrete areas

☐ Concrete cured a minimum of 3 days at above 50° F and in moist condition

☐ Curing compound, if used, was applied as soon as possible; coverage complete and of sufficient thickness to ensure adequate curing

☐ Forms stripped and removed after not less than 12 hours, depending on temperature; care taken not to damage fresh concrete

☐ Floors covered with straw or insulation blankets if night temperature below 40° F

Superintendent's Signature: _____ Date: _____

Interior Concrete Flatwork Checklist

Trade Contractor: _____   Job Name, Number: _____

Basement Slabs

☐ Work complies with local, state, and the national codes (even in areas where no local inspection exists)

☐ Plumbing installed, complete, and inspected

☐ Plumbing tested

☐ Soil treated for insect and pest control according to contract and code

☐ Floor drains installed so water drains into floor drain in prescribed areas, such as around hot water heaters with a slope of $\frac{1}{8}$ inch to $\frac{1}{4}$ inch per foot

☐ Area free from snow, ice, and debris; ground not frozen

☐ Gravel (where required) is in place, level ± 1 inch, depth specified ± 1 inch

☐ Vapor barrier place, if required

☐ No aluminum embedded in the concrete (very important)

☐ Floor concrete 3,000 psi

☐ Concrete mixed at least 20 minutes and not longer than 1 hour prior to placement

☐ Maximum slump 5 inches ± 1 inch

☐ No water sprinkled on surface to make troweling easier

☐ Slab level within ± $\frac{1}{4}$ inch in 10 feet and $\frac{1}{2}$ inch overall

☐ No premature finishing while bleed water present

☐ Forms stripped and removed after not less than 12 hours, depending on temperature

☐ Care taken to not damage fresh concrete

☐ Concrete protected from freezing by heating or insulation, as necessary

☐ Concrete cured for a minimum of 3 days at above 50° F and in moist condition

Superintendent's Signature: _____   Date: _____

Framing Checklist

Trade Contractor: _____   Job Name, Number: _____

- [ ] Work complies with local, state, and adopted national codes; areas where no local codes exist governed by state code regulations

Floor Support

- [ ] Posts properly secured and plumb in correct locations under bearing points
- [ ] Beams placed with camber up and with proper bearing
- [ ] Beam supports tight against beam
- [ ] Girder splices placed over bearing points
- [ ] Beams protected from moisture at concrete and masonry beam pockets and bear on non-compressible material
- [ ] Joists and subfloor at least 18 inches from exposed ground; wood girders at least 12 inches from exposed ground, or floor assembly constructed of natural decay-resistant or treated material

Floors

- [ ] Pressure-treated, redwood, or cedar sill plate material used against concrete or masonry
- [ ] Sill plates secured to foundation with j-bolts or anchors every 6 feet; each sill plate has at least 2 bolts or anchors and bolts or anchors are within 12 inches of each end of the plate
- [ ] Joist spans and layout according to plans
- [ ] Minimum bearing for joists $1\frac{1}{2}$ inches on wood and steel and 3 inches on masonry
- [ ] Minimum overlap of joists from opposite sides of a beam 3 inches
- [ ] Trimmer and header joists that span more than 4 feet are doubled
- [ ] Trimmer and header joists that span more than 6 feet are supported by hangers or supported on a ledger or bearing wall
- [ ] Tail joists over 12 feet are supported at their ends with hangers, ledgers, or bearing walls
- [ ] Notches should not exceed $\frac{1}{6}$ the depth of the joist; notches on the ends of joists can be $\frac{1}{4}$ the depth of the joist; no notches permitted in the middle third span of the joist
- [ ] Holes bored in joists not within 2 inches of the edges and the diameters do not exceed $\frac{1}{3}$ the depth of the joist
- [ ] Joist framing into the side of a wood girder supported by framing anchors or on ledger strips (minimum 2x2 material)
- [ ] Solid blocking installed over all bearing walls (not required in 1995 CABO code—check local codes)

☐ All square edges of subfloor blocked or nailed to joists

☐ Subfloor glued and nailed with 8d nails, 6 inches on center on edges and 12 inches on center at intermediate supports

Walls

☐ Natural decay-resistant or pressure-treated plates used against concrete or masonry

☐ Plans checked for proper stud size and spacing

☐ Top plates not lapped less than 48 inches; splices in the middle of studs

☐ Trimmers or jack studs fit tightly against headers

☐ Notches limited to 25 percent of the stud width for bearing walls and 40 percent for non-bearing walls

☐ Holes limited to 40 percent of the stud width for bearing walls and 60 percent for non-bearing walls

☐ Backing installed for drywall, cabinets, railings, door stops, and so forth

☐ Sheathing properly nailed, 6 inches on center around edges and 12 inches on center at intermediate supports

☐ Walls plumb and straight to within 1/8 inch in 8 feet

☐ Door rough openings plumb, square, and the proper sizes

☐ Firestopping installed at all concealed draft openings, both horizontal and vertical

☐ Walls with single top plates tied to intersecting walls with 1/8 inch x 1 1/2 inch metal straps or 2x lumber 16 inches long and fastened with two 16d nails on each side of the joint

☐ Braced wall lines installed in accordance with code

☐ Garage door trim installed and garage door track support in place

Roofs

☐ Rafters installed to plate with three 8d nails

☐ Trusses set so that peaks and vaults are aligned

☐ Truss bracing installed according to truss engineering details

☐ Gable end bracing installed and secured to a bearing wall

☐ Truss hangers flush to the bottom of trusses

☐ 2 x 4 strongbacks installed and bottom chords of trusses spaced correctly

☐ Roof sheathing secured with minimum 6d nails 6 inches on center around edges and 12 inches on center at intermediate supports

☐ Waferboard or oriented strand board (OSB) gapped 1/8 inch

☐ Barge rafters blocked or supported with lookouts every 48 inches on center

☐ Attic accesses installed with 12-inch side walls and 30-inch clearance above

☐ All fascia straight and secured at splices

☐ Firestop installed at each plate line

☐ Rafter and truss ties installed

Miscellaneous

Doors

☐ Verify door locations to within ± 1 inch

☐ Doors plumb to within ⅛ inch in 6 feet 8 inches

☐ Doors open and close easily and latch snugly against weather stripping

☐ A ⅛ inch reveal around the door and the jamb is uniform

Windows

☐ Verify window locations to within ± 1 inch and ± ⅛ inch in the kitchen

☐ Windows are installed level and square

☐ Windows open and close easily; windows latch securely

☐ Weepholes on bottom, drain out

☐ Rough windowsills for bedroom egress windows not more than 42½ inches from floor

Stairs

☐ Local code requirements checked

☐ Temporary guardrails and handrails installed

☐ Minimum headroom 6 feet 8 inches

☐ Maximum unit rise 8 inches (7¾ inches CABO)

☐ Minimum unit run 9 inches (10 inches CABO)

☐ Maximum variance between any two risers, or any two treads, is ⅜ inch

☐ Stair nosings extend past the risers about 1 inch

Decks

☐ Deck planks allow for water drainage and expansion of lumber

☐ Deck screws used to secure planks to structural supports

☐ Deck posts securely supported by posts anchors on concrete piles

☐ Deck stairs land on a concrete pad that is at least as wide and long as the width of the stairs

☐ Guardrails are 36 inches high and handrails between 34 inches and 38 inches off the nosing of the treads

☐ Handrails provide the proper hand grip

☐ Balusters no more than 4 inches apart

Chases

- ☐ Fire-blocking installed at roof lines and between floor levels
- ☐ Chases braced and anchored
- ☐ Chases provided with roof saddles or crickets on the roof's upside slope

Cleanup

- ☐ Lumber scraps put into dumpster or trash pile
- ☐ Extra lumber neatly stacked and covered
- ☐ Floors are swept clean of nails, sawdust, scraps, and trash

Other

- ☐ House observed from a distance and from all sides; structure appears correct, centered, straight, square, plumb, and level

Superintendent's Signature: _____ Date: _____

Mechanical and Electrical Checklist

Trade Contractor: _____ Job Name, Number: _____

Utility Laterals

☐ "Call Before You Dig" or "Blue Stakes" notified before digging; existing utilities properly marked (see chapter 1); call placed at least 48 hours before digging

Sewer

☐ Sewer lateral trenches dug so that pipe has a uniform slope of ¼ inch per foot; where ¼ inch per foot slope is impractical, the administrative authority may allow a slope of ⅛ inch per foot

☐ Trench sides sloped 1 foot out for every 2 feet vertical or a trench box is used to protect workers from collapsing trenches; when digging in gravel or sand and trench boxes are not used, sides are sloped 45 degrees

☐ Sewer lateral does not connect to any storm drainage system

☐ Sewer cleanout 4 inches, installed at connection of sewer line to the building drain

☐ Cleanout riser extends just above finish grade level

☐ Sewer lateral water- or air-pressure tested for leaks; a 10-foot head of water is used to develop 5 psi pressure in the pipe or a 5-psi air test is used; test remains in place for at least 15 minutes without any loss of pressure or head

☐ Inspection by local municipality required before pipe is covered over

☐ Trenches compacted in 12-inch lifts or water settled

Septic Tanks

☐ Private septic systems installed where no public or private sewer systems are provided (or if more than 200 feet away)

☐ Permits obtained

☐ Septic tank size determined by number of bedrooms in house

☐ Percolation test required to size the drain fields

☐ County officials consulted for specific installation requirements

Water

☐ Water laterals in common trenches with the sewer laid on a solid shelf of undisturbed soil or compacted fill to one side of the common trench and are at least 12 inches above the sewer line

☐ Water lines and sewer lines laid at same level are at least 10 feet apart

☐ Soldered joints not used underground; if underground joints required, use a flare fitting

☐ Water service lines caulked or sealed

☐ Sprinkling systems connected to culinary water include stop, waste, and backflow prevention devices

Electrical, Buried Lines

☐ Local electric utility consulted for specific requirements

☐ Trenches for electrical lateral lines dug a minimum of 26 inches deep

☐ Two-inch conduit used to protect electrical cable

☐ If the electrical utility installs the cable, the builder provides the conduit with a small rope or line inside to be used for pulling the cable through the conduit; the ends of the line are tied off to the panel and to the temporary utility post

☐ Ends of the underground conduit sealed

Electrical, Overhead Lines

☐ Local electric utility consulted for specific requirements

☐ Service-drop conductors not readily accessible

☐ Minimum clearance of service-drop conductors 3 feet from windows, doors, porches, and so forth

☐ Service conductors clear public roadways and alleys a minimum of 18 feet

☐ Conduit or raceways with rain-tight service heads used at meter masts

☐ Support cable used to sustain the weight of the service line; service lines installed with drip loops at the entrance of service heads

Gas

☐ Gas laterals installed by the local gas company

☐ If liquid propane gas is used, the homeowner has been consulted for the location of the LP gas tank (bottle)

☐ Installation of the gas lateral scheduled by builder with the gas company; a two-week notice is generally adequate

☐ Installation scheduled to follow foundation backfill

Superintendent's Signature: _____ Date: _____

Plumbing Checklist

Trade Contractor: _____ Job Name, Number: _____

☐ Work complies with local, state, and adopted national codes; areas where no local codes exist governed by state code regulations

☐ All parts of system visible for inspection

☐ All piping wrapped or otherwise protected when passing through concrete

☐ Notches not made in the middle third of the joist span; notches do not exceed $\frac{1}{6}$ the depth of the joist except at the ends, where notches may be $\frac{1}{4}$ the depth of the joist

☐ Holes in joists do not exceed $\frac{1}{3}$ the depth of the joist and are not within 2 inches of the top or bottom edges of the joist

☐ Holes bored through plates draftstopped with caulking or foam insulation

☐ Nail plates installed over studs, sole plates, and top plates where pipes are within $1\frac{1}{4}$ inches of the edge of the framing member

Water Supply

☐ Air test has 50 psi minimum for 15 minutes; air tests may be required by the local administrative authority to be 80 psi or more for 15 minutes (check local requirements)

☐ Water test has street pressure for 15 minutes

☐ No lead- or acid-bearing solders or fluxes used for potable water systems

☐ All cuts fully reamed

☐ Type M copper or galvanized pipe not used under slabs

☐ Soft copper tubing used underground

☐ Copper pipes supported every 6 feet

☐ Straps and hangers made of copper, copper plate, or plastic

☐ Main building water shutoff valve is a full-way valve

☐ Pressure reducing valves installed when the pressure at the meter exceeds 80 psi

☐ Temperature and pressure (T&P) relief valves and drains installed on water heaters

☐ Drains provided for pressure relief valves

☐ All hose bibs, valves, and pipe stub-outs to fixtures supported and secured

☐ Copper pipes and galvanized heating ducts have non-metallic separation

☐ Nail plates installed at studs, sole plates, and top plates where pipes are within $1\frac{1}{4}$ inches of the edge of the framing member

☐ Hose bibs (sill cocks) freeze-protected where installed in colder climates

Plastic Pipe Systems

☐ PEX (plastic) pipe supported every 4 feet

☐ No metal straps used with plastic pipe

☐ Minimum bend radius 12 times the nominal diameter of the tubing

☐ No plastic finish nipples or stub-outs

☐ A permanent sign with the words, "This building has non-metallic interior water piping" fastened to the main electrical panel

☐ Hose bibs secured so that torque on handles does not transmit to the plastic piping

Drainage, Waste, and Vent (DWV)

☐ Drainage and waste pipes have a uniform slope of ¼ inch per foot

☐ Vent pipes have a minimum slope of ¼ inch per foot

☐ Air test has 5 psi minimum for 15 minutes

☐ Water test has 10 foot head (5 psi) of water for 15 minutes

☐ All fixtures properly vented

☐ Fixtures set level, in alignment with adjacent walls, and with proper clearances

☐ Depth of shower dams and thresholds not less than 2 inches nor more than 9 inches

☐ Finished floor of shower receptor has a uniform slope from the sides toward the drain of between ¼ inch per foot and ½ inch per foot

☐ Showers have door openings for minimum 22-inch door

☐ Shower compartments greater than 1,024 square inches and capable of encompassing a 30-inch circle

☐ Trap arm lengths do not exceed the maximum distances based on slope of ¼ inch per foot:

Pipe Size	Maximum Trap Arm Length
1½ inches	3 feet 6 inches
2 inches	5 feet 0 inches
3 inches	6 feet 0 inches
4 inches	10 feet 0 inches

*Maximum length of water closet trap is 6 feet

☐ Trap vent at least two trap arm pipe diameters away from the trap weir

☐ Cleanouts accessible at the final inspection

☐ Cleanouts installed at the base of all stacks

☐ Change of direction on horizontal piping between cleanouts does not exceed 135 degrees

☐ Underground DWV not less than 2 inches

☐ Water closets have 3-inch drainage pipes and at least 2-inch vent pipes

☐ Maximum of three water closets on 3-inch horizontal branch drains

☐ Closet flanges secured to the floor with a fastener in each hole

☐ No S-traps

☐ No horizontal wet venting

☐ No wet venting across floors or between stories

☐ Vents connecting to horizontal drains are taken off above the center line of the drain served

☐ Condensate drains are provided near high efficiency furnaces, air conditioners, and heat pumps

Superintendent's Signature: _____   Date: _____

HVAC Checklist

Trade Contractor: _____   Job Name, Number: _____

☐ Work complies with local, state, and adopted national codes; areas where no local codes exist governed by state code regulations

☐ HVAC system sized to provide adequate supply of conditioned air to all heated and cooled spaces in the home

Heating, Cooling, and Ductwork

☐ Furnace has adequate clearance, typically 30 inches wide and 30 inches in front of combustion chamber (check local codes for exceptions)

☐ Type B vents have a minimum of 1 inch clearance from combustibles and electrical wiring

☐ Horizontal vent connectors sloped $\frac{1}{4}$ inch per foot

☐ Ductwork installed to maintain minimum ceiling heights of 7 feet 6 inches; 7-foot ceilings allowed in kitchen, bath, hall, and toilet areas

☐ Ductwork adequately supported

☐ Crimp joints of round duct have contact laps of at least $1\frac{1}{2}$ inches (dryer vent ducts excepted) and are secured with at least three sheet metal screws spaced evenly around the joint

☐ Flue and plenum chases fireblocked at all concealed floor and ceiling levels

☐ Safety pans with proper drains installed under all cooling units located in attics

☐ Duct pipe in the attic insulated

☐ Ductwork sealed

☐ Floor openings covered with wood or other temporary covers

☐ Combustion air grills are equipped with $\frac{1}{4}$-inch wire mesh

☐ Bathrooms having bathtubs or showers and no openable windows equipped with fans that vent to the exterior

☐ Evaporative cooler intakes 10 feet horizontally from or at least 3 feet below all flue or plumbing vent terminations

Gas

☐ Gas lines pressure tested under 10 psi pressure for 15 minutes

☐ Approved shut-off valves installed outside of and within 3 feet of appliances

☐ Gas piping adequately supported by metal straps or hooks; for rigid piping:

Pipe Size	Spacing of Supports
$\frac{1}{2}$ inch	6 feet
$\frac{3}{4}$ inch & 1 inch	8 feet
$1\frac{1}{4}$ inches	10 feet

☐ Each appliance connects with a ground joint union or an approved flexible connector

☐ Minimum ¾ inch gas piping for free-standing ranges

Superintendent's Signature: _____   Date: _____

Electrical Checklist

Trade Contractor: _____   Job Name, Number: _____

☐ Roof dried-in before electrical installed

☐ Work complies with local, state, and adopted national codes; areas where no local codes exist governed by state code regulations

☐ Services, conductors, raceways, boxes, and other electrical equipment adequately sized for intended use

☐ Telephone, communication, computer, antenna, security, and other wiring installed in proper locations

☐ Minimum one wall switch-controlled lighting outlet installed in all habitable rooms, bathrooms, hallway, stairways, garages, and at exterior entrances or exits

☐ Lighting outlets installed in attics, underfloor spaces, and utility rooms when these areas are used for storage and when containing equipment requiring servicing

☐ Non-recessed incandescent lighting fixtures more than 18 inches horizontally from any shelving in clothes closets; fluorescent lights more than 6 inches horizontally from closet shelving

☐ Boxes mounted ⅜ inch ± 1/16 inch out from the edge of the stud (wall face)

☐ No electrical equipment installed above gas meter

☐ All wiring protected from damage and exposed; wiring run in conduit

☐ Wiring staples used within 8 inches of plastic boxes, 12 inches of metal boxes, and every 4½ feet of run

☐ Receptacles not more than 12 feet apart

☐ All receptacles of the grounding type

☐ Receptacles installed at the ends of kitchen bars or countertops that exceed 6 feet in length

☐ At least two exterior receptacles installed, one in front and one in the rear of the house

☐ Receptacles and lighting outlets installed within 25 feet of mechanical equipment

☐ Boxes for heavy fixtures securely and adequately supported

☐ Rated or metal boxes used in walls separating houses from garages

☐ Holes for wiring drilled within 1¼ inches from the edge of studs protected with metal nail plates

☐ Penetrations through plates draftstopped with foam insulation or caulking

☐ Electrical meter sockets installed between 4 feet and 5 feet 6 inches from finished grade

☐ Small appliance receptacles 12 ga (20 amp) wire; includes at least two small appliance receptacles over countertops

☐ Grounding rod located at meter base and driven flush to the ground

☐ Electrical panel grounded to the copper water lateral; if plastic supply pipe used, secondary grounding is to another grounding rod or to the rebar in the concrete or block foundation

☐ Metal piping systems within the building bonded to the grounding system

☐ Ground-fault-circuit-interrupters (GFCIs) installed in bathrooms, garages, outdoors, crawl spaces, unfinished basements, all kitchen receptacles, and temporary construction receptacles

☐ Wires routed away from flues, attic access openings, metal gusset plates in trusses, and other areas where the wiring insulation could be damaged

☐ Venting fans installed in baths

☐ Breaker panels have clear working space to the floor that is 30 inches wide and 36 inches in front

☐ Panel breakers clearly marked and identified, with no abbreviations

☐ Corrosion inhibitor used for all aluminum connections

Superintendent's Signature: _____ Date: _____

Roofing Checklist

Trade Contractor: _____   Job Name, Number: _____

☐ All work complies with local, state, and the national codes (even in areas where no local inspection exists)

Asphalt or Fiberglass Shingles

Sheathing

☐ Sheathing completely nailed 8 inches on center

☐ Ends of all sheathing properly supported by rafters, trusses, or backing

☐ Attic ventilation fan cut out neatly with power saw and installed as indicated on the plans

Underlayment

☐ Asphalt saturated felt paper (15 #) covers entire roof surface within 24 hrs of sheathing completion

☐ Felt paper overlapped 2 inches (top lap)

☐ Felt paper overlapped 4 inches (end lap)

☐ Roof overhangs covered as required in specifications to prevent ice dam damage

☐ Ridge vents cut out to allow circulation

☐ Felt repaired as necessary; no roofing over missing felt (doing so voids the warranty on the shingles and violates the building code)

☐ All blisters in felt cut so felt will lie down

Drip Edge and Flashing

☐ Drip edge placed under felt paper at eaves and over the felt paper at gable ends

☐ Drip edge overlapped at least 4 inches

☐ Drip edge nailed securely to framing members at least 16 inches on center

☐ Flashing placed in and around all areas vulnerable to water leakage, especially around dormers and where the roof butts up against a wall

☐ Base and counter flashing placed around chimney

☐ Flashing (roof jacks or flanges) placed around all projecting objects on roof surface

☐ Flashing installed at valleys according to contract and specifications

☐ All flashing adjusted to slope of roof

☐ Black plastic cement used to seal off metal flashing in all necessary areas, such as around chimney and pipe flanges

☐ Crickets installed behind all chimneys and items projecting through the roof

☐ Skylight flashing and skylights installed according to manufacturers' recommendations

Shingles

☐ Color of shingles delivered to jobsite matches color specified in selections

☐ Shingles all from the same batch or factory run (spot-check run or batch number)

☐ Appropriate length (normally 1¼ inches), noncorrosive galvanized steel roofing nails were used to attach all shingles; nail shanks penetrate sheathing

☐ Starter course of shingles has tabs cut off so that the adhesive strips are at the outside edge of the roof; the adhesive strip is at the bottom edge of the roof so the first row of shingles seals properly

☐ Starter course overhangs 2 inches before rake mold installed

☐ Shingles attached starting at eaves and working upward

☐ Starter course and first course overlap eaves by ¾ inch and overlap gable ends by ½ inch

☐ Correct nailing patterns used to attach all shingles (normally 4 nails per shingle, up 5⅝ inches from edge of shingle)

☐ Nails perpendicular to the roof and snug against the shingle; nails that stick up will tear through the shingles or cause shingles to stick up and not seal properly

☐ Shingles installed with proper exposure (normally 5 inches)

☐ Shingle cut-outs staggered from row to row, according to manufacturer's recommendations

☐ Valleys woven or cut according to the builder's instructions

☐ Shingles, wrappers, and nails picked up and properly disposed of

☐ Roof checked to make sure all nails driven down and all shingles lay flat

☐ Scraps and loose nails cleaned off roof and ground (walk all the way around the house)

☐ Scraps and trash placed in designated trash site; clean up all wrappers and scraps and place in dumpster or designated area

☐ Leftover material stacked neatly on a pallet ready to return for credit

Wood Shingles and Shakes

☐ Proper grade (for example, Number 1, Blue Label) and length (16 inch) shingles used

☐ Double starter course on first row of shakes at the eaves

☐ Nails long enough to go through the shingle or shake and penetrate the sheathing at least ½ inch

☐ Nails driven flush with surface of shake

☐ 18-inch-wide strip of 15 # felt placed 18 inches over top portion of each shingle or shake, extending onto the sheathing according to manufacturer's specifications

☐ Shingles overhang 1 inch to 1½ inches over the eaves and 1 inch over all rakes

☐ Two nails per shingle or shake

☐ Spacing between shakes or shingles at least ½ inch apart and offset at least 1½ inches from adjacent courses

☐ Appropriate exposure for each shake or shingle:

       16-inch shingle or shake:
       3¾ inches for slope < 4/12 or 5 inches for slope > 4/12

       18-inch shingle or shake:
       4¼ inches for slope < 4/12 or 5½ inches for slope > 4/12

       24-inch shingle or shake:
       5¾ inches for slope < 4/12 or 5¾ inches for slope > 4/12

☐ Hips and ridges properly capped

Tile Roofing

☐ # 30 asphalt-saturated felt paper under tiles

☐ Head lap 2 inches and side lap 6 inches

☐ Battens 1x2s with a ½-inch break every 4 feet 0 inches; except first row of battens 2x2s to provide for proper slope on starter strip of tiles

☐ Battens placed 13 inches on center for 16½-inch tiles

☐ Flashing around vents in roof ; flashing placed 3 inches under the tile on top

☐ Tiles correctly installed and interwoven between each tile

Superintendent's Signature: _____ Date: _____

Siding Checklist

Trade Contractor: _____   Job Name, Number: _____

Siding Selection: _____   Color: _____

Vinyl or Aluminum Siding

- ☐ Correct brand
- ☐ Correct color
- ☐ Correct style
- ☐ Correct finish
- ☐ Final grade (soil) is at least 8 inches below the bottom of the siding
- ☐ Siding overhangs the foundation wall by 1½ inches
- ☐ Flashing and drip caps installed over all door and window openings
- ☐ Siding nailed 16 inches on center
- ☐ Siding nails on vinyl siding loose enough to allow siding panels to move from side to side with relative ease
- ☐ Aluminum siding nailed with aluminum nails
- ☐ Aluminum siding grounded
- ☐ Panels run straight level lines, corner to corner; inside and outside corners line up with each other
- ☐ Joints lapped to overlap away from the greatest traffic area
- ☐ Roof flashed in all necessary areas when siding installation precedes installation of roof shingles

Wood Siding

- ☐ Back side of the siding backed and/or primed before nailing
- ☐ Siding extends 1½ inches below top of foundation wall
- ☐ Bevel, channel, rustic, and similar lap sidings overlap 1 inch
- ☐ Noncorrosive rust-resistant nails used for all wood siding
- ☐ Nails appropriate size for the siding being used
- ☐ Siding nails 12 inches to 16 inches on center, driven into solid material (plywood or OSB)
- ☐ Spacer strip placed under starter course of bevel siding
- ☐ Tongue-and-groove siding nailed on top edge only and at 45-degree angle from top edges to the tongue
- ☐ When 5/4"x4" or 5/4"x6" corner board and foam sheathing material are used, make sure that the ends of all horizontal siding is nailed to solid backing material
- ☐ Caulk between the siding and the corner 1"x 4", inside corners and outside corners

☐ Solid sheathing only under shingle and shake siding

☐ Shingle and shake siding has double shingle or shake for starter strip

☐ Each shingle or shake has at least two nails

☐ Shingles and shakes have proper exposure for the thickness grade and species of shingle or shake

Shingle Length	Exposure
16 inches	7½ inches
18 inches	8½ inches

Shingle Length	Exposure
18 inches	8½ inches
24 inches	11½ inches
32 inches	15 inches

☐ Check siding for splits, cracks, and knot holes

Cornice

☐ Gable end overhangs have solid soffit material; all other overhangs open to attic perforated

☐ Wood fascia 1x or 2x material face-nailed to each truss or rafter

☐ Fascia straight and level ± ⅛ inch

☐ Screen soffit vents installed as required in soffit between trusses or rafters

☐ Frieze board and molding installed as required

Shutters

☐ Correct style

☐ Correct size

☐ Correct color

☐ Correct location

☐ Master mounts properly located for exterior light receptacle and dryer vents

Superintendent's Signature: _____ Date: _____

Brick Checklist

Trade Contractor: _____   Job Name, Number: _____

Brick Selection: _____   Mortar Selection: _____

☐ Work complies with local, state, and the national codes

☐ Brick and mortar agree with color selection

☐ Brick was not laid wet

☐ Wall ties nailed on before laying brick at 16-inch-on-center vertical and 16-inch-on-center horizontal intervals

☐ Felt vapor barrier (15 #) installed as required between the brick and plywood or OSB sheathing

☐ Brick installed plumb, level, and straight ± 1/8 inch

☐ Brick installed with uniform head and bed joints

☐ Bricks properly aligned so that no brick sticks out past others

☐ Bed and head joints full

☐ Mortar joints tooled or raked according to specifications; brick kept clean

☐ Raked joints not more than 3/8 inch deep

☐ Weepholes clean, open, and spaced according to specifications

☐ Chimneys properly reinforced and tied to building according to code (especially important in seismic zones 3 and 4)

☐ Brick mortar cleaned off brick and tags swept clean

☐ Brick cleaned with proper solution (typically 10 percent solution of muriatic acid, depending on brick)

☐ Proper protective measures taken to prevent mortar from freezing

☐ Mortar cleaned off all windows, doors siding, roof shingles, and soffits

☐ Trash cleaned up and properly disposed of; broken brick, bands, pallets, cement, and mortar bags are put in trash bin or piled in one location as indicated by the superintendent

Superintendent's Signature: _____   Date: _____

Gutters and Downspouts Checklist

Trade Contractor: _____   Job Name, Number: _____

Size: _____   Color: _____

- ☐ Correct color and size as stated on work order
- ☐ Gutters have correct fall towards downspout
- ☐ Short connecting downspouts installed square with house frame
- ☐ Downspouts installed to correct height from ground (8 inches to 10 inches off dirt)
- ☐ If decks interfere with a downspout, downspout turned and installed down end of house corner
- ☐ Downspouts securely connected and in place with no dents; explain any exceptions: _____
- ☐ Drop downspouts on open side of retaining walls
- ☐ If porch is not on, downspout left with straps on house roof for deck crew to install
- ☐ Gutters nailed to every rafter or to truss ends, not to fascia boards
- ☐ Gutters installed tight to fascia or siding at raised roofs to prevent runoff passing end of gutter
- ☐ Trash cleaned up and properly disposed of

Superintendent's Signature: _____   Date: _____

Insulation Checklist

Trade Contractor: _____   Job Name, Number: _____

Foam, Caulking, and Baffles

☐ Work complies with local, state, and the national codes and the Model Energy Code (even in areas where no local inspection exists)

☐ Sill sealer insulation or caulking installed at foundation sill plate

☐ Foam or caulking installed at bottom plates, exterior corners, and between double studs

☐ Gaps around windows insulated, foamed, or caulked as needed

☐ Gaps around exterior doors and windows insulated, foamed, or caulked

☐ Foam insulation does not create pressure against doors and windows (causing doors and windows to stick or bind)

☐ All pipes and wired penetrating outlet boxes on outside walls caulked or insulated

☐ Foam installed where all pipes and wires pass through top and bottom plates

☐ Ceiling insulation baffles installed

☐ No gaps visible at baffle corners

☐ Baffles extend all the way to the outside of the double top plate

☐ Baffles have 1 inch space at top

☐ Baffles fit properly, allow maximum insulation depth over plates

☐ Baffles cut neatly around vent stacks

☐ Baffles dry and in good condition

Walls

☐ Holes in insulative sheathing patched with foil tape (no daylight visible)

☐ Insulation installed at rim or header joist

☐ Batts installed in cantilevered floor system

☐ Batts neatly fitted without compression

☐ Batts contact top and bottom plates, snugly fill entire stud cavity

☐ R-Values as specified

☐ Insulation tightly cut to fit irregular spaces

☐ Insulation installed behind tubs, showers, and around HVAC closet (double-check; extremely important)

☐ Insulation cut out at outlets, switch boxes; cut out piece placed behind the electrical box

☐ Insulation split and fitted around wires without compression

☐ Fireplace chase insulated as required

☐ Insulation installed around metal fireplaces as required

☐ Vapor barrier installed properly; no holes, splices

☐ Vapor barrier has perm rating of 10 or less

☐ Vapor barrier proper thickness, overlapped

☐ Tray ceiling insulated with batts equal to ceiling R-Value

☐ Walls insulated behind corner tub before the tub is installed

Ceilings

☐ Ceiling access panel insulated with batt insulation to same R-Value as ceiling, as specified

☐ Vaulted stairwell ceilings insulated to same R-Value as rest of ceilings

☐ Vaulted ceilings insulated to same R-Value as the rest of the ceiling

☐ Ceiling insulation blown to an even depth, not over-blown into ventilation space of ceiling baffles

☐ Insulation certificate installed in the attic

Crawl Space

☐ Batts installed in crawl space to R-Values as specified

☐ Batts installed in cantilevered floor system

☐ Batts installed in the same plane to butt up neatly together without gaps

☐ Wire mesh (18 gauge) secured to underside of joists; or 50 lb. to 60 lb. test plastic twine nailed 12 to 18 inches on center; or tiger teeth stays installed every 24 inches to hold insulation in place

☐ Support system strong and continuous

☐ Foundation vents not blocked

☐ Vapor barrier or floor insulation batts placed above the insulation (on the warm side of the insulation)

Cleanup

☐ All trash and insulation bags hauled off

☐ R-Values as specified

Superintendent's Signature: _____   Date: _____

Drywall Checklist

Trade Contractor: _____   Job Name, Number: _____

Hanging

☐ Work complies with local, state, and the national codes (even in areas where no local inspection exists)

☐ Delivery-related damage to doors or windows duly noted

☐ Nail guards installed to protect wiring and plumbing from intruding nails and screws

☐ Proper drywall thickness and type used in all places

☐ Type X drywall used on garage firewalls, ceilings, and underneath all stairs, as required by code

☐ Moisture-resistant drywall installed in bathroom around tubs and showers as specified

☐ Joints tight (especially vaulted ceilings) ± $\frac{1}{8}$ inch

☐ Electrical boxes cut out neatly, without gaps ± $\frac{1}{8}$ inch

☐ Fireplace face covered, where applicable

☐ Drywall adhesive used on all interior stud walls; nails and screws set while adhesive still wet so board bonds with the glue

☐ Nails on ceiling spaced 7 inches on center maximum

☐ Nails on walls spaced 8 inches on center maximum

☐ Screws spaced 12 inches on center maximum

☐ Windows and corners wrapped with metal corner bead or J-mold as required and as specified on the prints

☐ Corner bead straight and nailed at least 8 inches on center on both sides; corner bead corners butt tightly together

☐ Leftover drywall removed from the house immediately after completing hanging and placed in designated dumpster or trash pile as instructed by the superintendent (this is a safety as well as a housekeeping issue)

☐ Drywall not left outside windows or in front yard, even for one night; violators of this policy backcharged fully for inconvenience and cleanup expense

☐ Floors swept and broom-clean after hanging

Finishing

☐ Walls have smooth finish everywhere

☐ Corners have smooth finish everywhere

☐ No loose nails, nail pops (check randomly)

☐ Finish fits properly around windows

☐ Finish clean, fits properly around outlet boxes

☐ Joints straight and clean

- ☐ No bubbles or gaps in tape
- ☐ Corners tight (so they will not crack when base is nailed)
- ☐ Metal corners bead-covered with mud, sharp and neat
- ☐ Drywall mud applied all the way to bottom on inside corners and at windows
- ☐ Drywall corners finished smoothly and completely
- ☐ Corners sanded smooth
- ☐ Ceilings have uniform texture and completely cover the ceiling area; no hollow spots
- ☐ Floors scraped and swept; all dirt and debris removed to the designated location
- ☐ Mud cleaned off doors, tubs, showers, fireplace face, and any other surface
- ☐ All trash cleaned up and properly disposed of

Superintendent's Signature: _____ Date: _____

Interior Trim Checklist

Trade Contractor: _____   Job Name, Number: _____

☐ Work complies with local, state, and the national codes (even in areas where no local inspection exists)

☐ Shims installed behind all exterior doors and hinge screws installed (The siding trade contractor or the framer normally tacks the exterior doors in place through the brick mold; the trim carpenter installs the shims and hinge screws.)

Set Interior doors

☐ Set doors in carpet areas up $1\frac{1}{2}$ inches off carpeted and wood floors, $\frac{1}{2}$ inch off vinyl floors, $\frac{3}{4}$ inch off basement floors

☐ Hinge side of doors plumb $\pm$ $\frac{1}{8}$ inch, nailed tightly against jack

☐ Shim hinge side only if opening out of plumb or for special doors such as in hallways (to center them in the opening)

☐ Shim knob side at top, bottom, and at strike plate

☐ Doors operate properly, quietly

☐ Maintain $\frac{1}{8}$ inch margin or spacing between door and doorjamb to allow for paint and temperature and humidity changes so door won't bind or stick

☐ All miter joints glued, fit tightly; top and edge of casing nailed through miter joint to hold miter together (face nails in casing at miter joints insufficient to ensure joint won't open up as the house settles)

☐ Face nails held back about 3 inches to 4 inches from ends of the casing so casing is not split

☐ No gaps on 45 degree miter joints on door casings, joints are flush

☐ Casing nailed to the jamb and to the jack studs 16 inches on center, nails are set

☐ Locks installed on all doors

☐ Door bumpers installed on all doors

Thresholds

☐ Adjust threshold on exterior doors (just snug with no daylight)

☐ $\frac{3}{4}$-inch board and threshold installed on furnace room

Run Baseboards

☐ Baseboard raised $\frac{1}{2}$ inch in carpet areas

☐ Baseboard raised $\frac{3}{8}$ inch in vinyl areas on $\frac{3}{4}$ inch tongue and groove

☐ Baseboard cut and tacked in place in vinyl areas

☐ Baseboard nailed with two number 8s on each stud only and within 8 inches of all corners

- [ ] Inside corners mitered, coped for tight fit
- [ ] If cabinets not installed, run all base and tack in place
- [ ] Remove all shims from under baseboards

Fireplace

- [ ] Correct mantel style
- [ ] Trim installed next to fireplace if gap exists between fireplace and wall
- [ ] Mantel installed securely
- [ ] Crown molding mantel runs on brick faces only

Planter Shelf

- [ ] Planter shelf capped with _ inch material big enough to trim underside with case molding (use _ inch plywood if necessary)

Trim Exterior

- [ ] Exterior entrance door shimmed and trimmed
- [ ] Dead bolts installed on exterior entrance doors

Windows

- [ ] Windowsills installed, level, and secure
- [ ] Sills tight against drywall

Fan Block

- [ ] Paddle fan block with 4 inches of # 8 wood screws

Disappearing Stairs

- [ ] Disappearing stairways to attic installed with 3 inch screws after nailing frame and trim
- [ ] Legs cut on proper diagonal, ensure maximum support for stairs
- [ ] Left so homeowner can pull out and reinstall

Cleanup

- [ ] Excess material neatly stacked for return or use on another job
- [ ] Trash and unusable material removed and placed in the designated area
- [ ] House thoroughly swept out and cleaned; all dirt removed to designated trash area
- [ ] Locks adjusted (double-check; readjust if necessary)

Superintendent's Signature: _____   Date: _____

Painting Checklist

Trade Contractor: _____ Job Name, Number: _____

Paint Order

☐ Correct type, color (per color selection sheet) specified on order
☐ Correct type, color, and sheen (double-check on delivery to jobsite)

Interior Trim Paint

☐ Edges of windowsills caulked with white caulk
☐ Tops of all baseboards caulked
☐ Nail holes filled with painter's putty

Interior Walls

☐ Damaged drywall repaired and painted
☐ Walls double rolled
☐ Countertops and tubs in bathrooms caulked
☐ Edges of windows cleaned and caulked where they meet the drywall
☐ No roller marks on walls

Basement

☐ Stairway painted
☐ Handrail finished as specified

Exterior Paint

☐ Rake molding caulked where it meets the eave
☐ Edges of windows caulked
☐ Nail holes in rails filled and painted
☐ Exterior raw wood has one coat of primer and topcoat of paint, as specified
☐ No paint on porch floor, concrete drives, garage floor, and so forth

Punch Out

☐ No paint on sweeps at bottoms of exterior doors
☐ Touch up all baseboard, doors, and trim as necessary
☐ Touch up all walls with a roller as necessary
☐ Repaint outside of steel doors as necessary
☐ Remove paint and stain from tubs, cabinets, or vanities

☐ Cleanup (daily)

☐ Store flammable materials in a fireproof cabinet or remove from jobsite each day

Superintendent's Signature: _____   Date: _____

Mirror and Accessory Checklist

Trade Contractor: _____   Job Name, Number: _____

Mirror Size: _____   Accessory Color: _____

☐ Correct colors (per selection sheet)

☐ Accessories match room fixtures

☐ Mirrors and accessories all in place

☐ No chips, scratches, or cracks

☐ Mirror clips mounted into studs

☐ Accessories level and plumb ± $\frac{1}{16}$ inch

☐ Thin set has been wiped clean

☐ Mirror height standard 40 inches unless otherwise specified or requested

☐ Mirror width 2 inches less than lavatory tops when open on one end, 4 inches less when wall-to-wall measurements are used

☐ Trash and boxes cleaned up and put in designated location or hauled off

Superintendent's Signature: _____   Date: _____

Final Checklist

Trade Contractor: _____   Job Name, Number: _____

Pre-Closeout Inspection

☐ Work complies with local, state, and the national codes (even in areas where no local inspection exists)

Exterior

☐ Stucco correctly installed

☐ 5 percent grade away from house

☐ Final grade smooth, backfill properly settled

☐ Sidewalks installed according to plan

☐ Loose concrete cleaned up, hauled off

☐ Steps slope away from home

☐ Hose bibs installed securely, shut off properly

☐ No broken, cracked, or scratched windows

☐ No blown-off shingles; no popped nails on shingles

☐ Garage door opens easily, does not stick

☐ Brick cleaned

☐ Gutters and downspouts properly installed

☐ Exterior trim properly installed

☐ Exterior paint completed, masking cleaned up

☐ Shutters securely installed

☐ Roof vents correctly installed and operational

☐ Chimney cap correctly installed

☐ Flashing and counterflashing correctly installed

☐ Heating vent caps installed

☐ Roof jacks caulked or sealed with plastic cement

☐ Windows caulked, including top

☐ Screens installed

☐ Siding properly nailed

☐ Trim properly nailed and caulked

☐ Deck properly installed, nailed, and painted

☐ Steps and railings secured

☐ Porches and decks swept

☐ Garage doors properly installed and operating

☐ Garage floors cleaned

- [ ] General site cleanup
- [ ] Thresholds on entrance doors cleaned, tape removed
- [ ] Thresholds adjusted
- [ ] Exterior locks checked, function properly

Interior

- [ ] Basement floor scraped and swept; floor washed if necessary
- [ ] All doors open freely, do not close by themselves
- [ ] All doors clear the carpet or other floor coverings
- [ ] Door bumpers installed
- [ ] Base joints fit tightly in corners and against the wall
- [ ] All shelving installed
- [ ] Drywall in garage hung and taped (if required)
- [ ] Check wall for chips or scratches
- [ ] Adequate insulation properly installed in attic
- [ ] Attic access properly installed and insulated
- [ ] Outlets and switches installed leaving no holes in drywall
- [ ] Light fixtures have bulbs and function (test)
- [ ] Outlets work (test)
- [ ] Fans work (test)
- [ ] Garbage disposal works (test)
- [ ] Dishwasher properly connected, functions (test)
- [ ] Oven and stove function (test)
- [ ] All electrical trim in place
- [ ] All smoke alarms function (test)
- [ ] Check water heater; make absolutely sure it is full of water before turning on
- [ ] Check all faucets and drains to make sure they work properly (test)
- [ ] Plumbing fixtures properly installed; turn on water in all sinks to assure function
- [ ] Hot water tested at all sinks
- [ ] Check for leaks
- [ ] Toilets flush (test)
- [ ] Floor drains have water in traps
- [ ] Bath hardware securely installed
- [ ] Sinks, commodes, and tubs well cleaned
- [ ] Drywall mud removed from top and side of tub units
- [ ] All stain or paint removed from bottoms of tubs at shoe mold
- [ ] Tubs, cabinets, and countertops checked for chips and scratches; superintendent notified if damaged

- [ ] Mirrors securely installed
- [ ] Mirrors cleaned in baths
- [ ] Shower rods and doors installed and caulked
- [ ] Tile installed, cleaned, and caulked
- [ ] All heat diffusers (registers) in place
- [ ] Filter in furnace
- [ ] Furnace tested, turned on (in winter)
- [ ] Air conditioning tested, turned on (in summer)
- [ ] Thermostat set at 70°F
- [ ] Appliances installed, working properly
- [ ] Cabinets properly installed
- [ ] Cabinet doors and drawers operate properly
- [ ] Cabinet trim properly installed
- [ ] Countertops installed and properly caulked where needed
- [ ] Open all cabinets and eliminate squeaks
- [ ] All drawers work properly
- [ ] Cabinet interiors vacuumed out
- [ ] Cabinet exteriors wiped down with polish
- [ ] Any glue removed from countertops
- [ ] Rails and balusters securely fastened and stained
- [ ] All the windows tested to make sure they open without binding
- [ ] Bottom window sash removed and tracks cleaned out
- [ ] All drywall mud removed from window frames
- [ ] Windows and doors cleaned with no streaks
- [ ] Bottom track in patio doors cleaned out
- [ ] All parts of the house appear clean and free of construction debris
- [ ] All construction debris cleaned up and properly disposed of

Superintendent's Signature: _____ Date: _____

Homeowner's Signature: _____ Date: _____

Homeowner's Signature: _____ Date: _____

Additional Resources

Accounting and Financial Management for Builders and Remodelers, by Emma Shinn
(Home Builder Press)

Bar Chart Scheduling for Home Builders, by Thomas A. Love
(Home Builder Press)

Contracts and Liability for Home Builders, Remodelers, and Developers, by David Jaffe
(Home Builder Press)

Contracts with the Trades: Scope of Work Models for Home Builders, by John Fredley
and John Schaufelberger
(Home Builder Press)

Customer Relations Handbook for Builders, by Carol Smith
(Home Builder Press)

*Destination Quality: How Our Small-Volume Building Firm Used TQM to Improve
Our Business,* by Gilbert Veconi with Charles A. Layne
(Home Builder Press)

How to Hire and Supervise Subcontractors, by Bob Whitten
(Home Builder Press)

Job Descriptions for the Home Building Industry
(NAHB Business Management Committee)

Production Checklist for Builders and Superintendents
(Home Builder Press)

Production Manual Template
(NAHB Builder Business Services)

Quality Management: Best Practices for Home Builders, edited by Edward Caldeira
(Home Builder Press and the NAHB Research Center, Inc.)

The Model Safety & Health Program for the Building Industry, 2nd Edition, (NAHB
    Labor, Safety and Health Services Department)

Scheduling Residential Construction for Builders and Remodelers, by Thomas A. Love
    (Home Builder Press)

To order any of these books, to request an updated catalog of Home Builder Press
titles, or for more information on any NAHB publications, write or call:

Home Builder Bookstore®
1201 15th Street, NW
Washington, DC 20005-2800
(800) 223-2665
www.builderbooks.com